Advanced BASIC Meta-Analysis

Advanced BASIC Meta-Analysis

Brian Mullen
Syracuse University

LAWRENCE ERLBAUM ASSOCIATES, PUBLISHERS
1989 Hillsdale, New Jersey Hove and London

Lawrence Erlbaum Associates, Inc., Publishers
365 Broadway
Hillsdale, New Jersey 07642

Library of Congress Cataloging-in-Publication Data
Mullen, Brian.
Advanced BASIC meta-analysis / Brian Mullen.
p. cm.
Bibliography: p.
Includes index.
ISBN 0-8058-0502-8
1. Mathematical analysis—Data processing. 2. BASIC (Computer program language) 3. Meta-analysis. I. Title.
QA276.4.M84 1990
515'.028'55133—dc20 89-7855
CIP

Printed in the United States of America
10 9 8 7 6 5 4 3 2

To Bob

Contents

Preface

This book presents an overview of the strategies, techniques, and procedures of meta-analysis. Meta-analysis generally refers to the statistical integration of the results of independent studies. This quantitative approach to the assessment of research domains represents a growing and compelling emphasis on rigor and precision in the summarization and integration of research literatures.

I have had three goals in writing this monograph. First, I have attempted to provide a clear and user-friendly introduction to the procedures and rules of effective meta-analytic integration. Second, I have attempted to present the implicit assumptions and strategies that guide successful meta-analytic integrations. Third, I have developed a meta-analytic database management system that allows the user to create, modify, and update a data base, including the relevant statistical information and predictors for a given research domain. This software system allows the user to perform a full complement of meta-analytic statistics with the speed and flexibility of a database management system. In addition, this software system constructs a wide variety of meta-analytic graphics displays.

In presenting the procedures and rules of meta-analysis, this book may serve as a useful introduction to the quantitative assessment of research domains. The techniques that figure most prominantly in this presentation are those that were developed and popularized by Rosenthal and Rubin (e.g., Rosenthal, 1968, 1979, 1983, 1984; Rosenthal & Rubin, 1971, 1978, 1986), most of which can be traced back to the seminal "meta-analytic" efforts of Fisher (1932, 1938), Pearson (1933a, 1933b), Snedecor (1946), and Thorndike (1933). In addition, techniques suggested by Hedges and Olkin (1983), Glass (1980), and Light and Pillemer (1984) are incorporated in this presentation. In detailing the implicit assumptions and strategies, this book may serve as a valuable sourcebook for those who have already conducted meta-analyses. Some of the techniques and strategies detailed herein may suggest fruitful new approaches to the experienced reviewer. Finally, the meta-analytic data base management system should enhance the practicability of these techniques for the novice and the experienced reviewer alike.

If you have glanced at the table of contents, you will have noticed that meta-analytic statistics per se are not presented until halfway through the book. This is quite intentional and quite realistic. No matter how important the statistical techniques themselves might be, their application to an ill-conceived problem and to a poorly derived database would be worthless. The meta-analytic statistics are introduced after the necessary groundwork has been laid.

ACKNOWLEDGMENTS

Many people have provided helpful comments on earlier drafts of this book. A debt of gratitude is owed to Alice Eagly, Jack Gleason, Eric Gold, Li-tze Hu, Blair Johnson, Craig Johnson, Eduardo Salas, and Colleen Smith for helpful feedback provided at various stages of this work. In addition, Eric Gold and Blair Johnson have both gone beyond the call of duty in helping me to fine-tune and compile the software. I am especially grateful to Judi Suben and Hollis Heimbouch of Lawrence Erlbaum Associates, both of whom were instrumental in tranforming this project from idea into reality. Finally, I express my appreciation to Robert Rosenthal, to whom this book is dedicated: for always being on the other end of the phone with sound advice, cheering encouragement, and innovative solutions to the complex puzzles of meta-analysis.

1

Meta-Analysis: An Introduction

Scientists are not dependent on the ideas of a single man, but on the combined wisdom of thousands of men, all thinking the same problem, and each doing his little bit to add to the great structure of knowledge which is gradually being erected.

—Lord Rutherford (1933)

Meta-analytic statistics provide a single set of numbers that describe and summarize the results of independent pieces of research. Meta-analysis generally refers to the statistical integration of the results of independent studies. This quantitative approach to the assessment of research domains represents a growing trend toward more rigorous and more precise summarization and integration of research literatures. Meta-analysis does not describe a single statistical procedure that distills a domain of research into one simple answer. Rather, meta-analysis embodies a constellation of different statistical techniques, developed and suited for specific purposes, and a general conceptual approach to the problem of summarizing, integrating, and testing theoretical issues with the results of previous research.

In coining the term *meta analysis,* Gene Glass identified three levels of analysis, and thereby established a meaningful context for meta-analytic statistical procedures. *Primary analysis* refers to the original statistical analysis of data, as it is collected, by the researcher who collected the data. In other words, if you conducted a study on the effects of X on Y, you might collect measurements of X and Y and calculate a correlation coefficient between these sets of measures; alternatively, you might calculate a t test, comparing the mean Y obtained under conditions of low X with the mean Y obtained under conditions of high X. These instances would constitute primary analyses. *Secondary analysis* refers to analysis of data by someone other than the researcher who collects the data, possibly for purposes, or with analytic strategies, other than those of the original researcher. For example, if I am really interested in gender differences, and if you are gracious enough to make your original data available to me, I might reanalyze your data by calculating an analysis of variance contrasting the effects of X on Y in your male subjects with the effects of X on Y in your female subjects. This would constitute a secondary analysis. Finally, *meta-analysis* refers to analysis of the results of several independent studies. For example, if you have conducted one study on the effects of X on Y, and I have conducted a second study of the effects of X on Y, and 20 other investigators have conducted studies of the effect of X on Y, a meta-analysis would provide a numerical summary and integration of the results of these separate studies. This meta-analysis would include such things as an index of the combined probability of the obtained results, an average magnitude of effect, a gauge of the heterogeneity of significance levels and effect sizes, an indication of the extent to which study outcomes can be accurately predicted by theoretically interesting study characteristics, and so on. In a sense, then, meta-analysis refers to methods of statistical analysis wherein the units of analysis are the results of independent studies, rather than the responses of individual subjects.

In passing, we might note that some authors have taken the extension of higher levels of analysis to its illogical conclusion, in proposing the strategies of "mega-analysis" (Dillon, 1982; Smith, 1982), "meta-meta-analysis" (Kazdin, Durac, & Agteros, 1979), and "super-analysis" (Dillon, 1982), all of which variously refer to the statistical integration of the results of independent meta-analyses! Although this notion is intriguing, the statistical techniques of meta-analysis were developed for databases where the units of analysis are statistical tests of a hypothesis at the primary level. No one has developed the statistical techniques that could be defensibly applied to databases where the units of analysis are, themselves, meta-analytic statistics. In the present context, we restrict ourselves to the less ambitious goal of integrating the results of separate primary-level studies from the perspective of meta-analysis.

Meta-analysis has become increasingly popular over the past decade and has been applied to a broad spectrum of phenomena in the social sciences. A partial list of phenomena that have been examined using meta-analytic techniques includes: interpersonal expectancy effects (Rosenthal, 1968, 1969; Rosenthal & Rubin, 1971, 1978); characteristics of volunteers for behavioral research (Rosenthal & Rosnow, 1975); the effectiveness of psychotherapy (Smith & Glass, 1977); the effects of class size on attitudes and achievement (Glass & Smith, 1979); characteristics of deceptive communication (DePaulo, Zuckerman, & Rosenthal, 1980); the effects of hand-held calculators in precollege mathematics education (Hembree & Dessart, 1986); gender differences in social influence (Eagly & Carli, 1981); effects of desegregation (Israel, 1981; Miller & Davidson-Podgorny, 1987); differential effects of exercise on serum lipid and lipoprotein levels (Tran & Weltman, 1985); effects of computer-assisted instruction (Bracey, 1982); the effectiveness of tutoring (Cohen, Kulik, & Kulik, 1982); teacher expectancy effects (Dusek & Joseph, 1983); the relation between problem-solving skills and adjustment (Almeida & Denham, 1984); the determinants of successful salesperson performance (Churchill, Ford, Hartley, & Walker, 1985); the effects of source strength and immediacy in group contexts (Mullen, 1985); the relative efficacy of avoidant and nonavoidant coping strategies (Suls & Fletcher, 1985); gender differences in helping behavior (Eagly & Crowley, 1986); gender differences in aggressive behavior (Eagly & Steffen, 1986); the validity of employment interviews (Arvey & Campion, 1982); the relation between negative mood and helping behavior (Carlson & Miller, 1987); social projection effects (Mullen & Hu, 1988); the illusory correlation in stereotyping (Mullen & Johnson, 1989); the effectiveness of brainstorming in groups (Mullen, Johnson, & Salas, 1989); The prevention of venous thromboembolism in general surgical patients (Clagett & Reisch, 1980); cardiac rehabilitation after myocardial infarction (Oldridge, Gvyett, Fisher, & Rimm, 1988); and, the relation between alcohol consumption and risk of breast cancer (Longnecked, Berlin, Orza, & Chalmers, 1988).

Something that many novices at meta-analysis fail to recognize is that the fundamental issues, the basic approach, and many of the techniques of meta-analytic statistics have been around for a very long time. Excellent brief histories of meta analysis have been provided by R. Rosenthal (1984) and Bangert-Drowns (1986). Therein, several early examples of meta-analytic techniques and approaches are illustrated, including Lush's (1931) derivation of mean effect sizes, Snedecor's (1946) assessment of the heterogeneity of a set of effect sizes, Fisher's (1932, 1938) and Pearson's (1933a, 1933b) combination of probability levels, and Thorndike's (1933) prediction of variability in effect size as a function of study characteristics. Indeed, Rosenthal's (1961, 1963, 1964,

1966) own early efforts to summarize and integrate the results of research on experimenter expectancy effects comprise one of the earliest and most comprehensive efforts to summarize and integrate the significance and magnitude of the results of separate studies in social science. It was not until Gene Glass's (1976) elegant labeling of this conceptual and analytic perspective as *meta-analysis,* and his controversial integration of research on psychotherapy outcome (Smith & Glass, 1977), that this approach achieved the currency and/or the notoriety that it enjoys today.

In a preliminary effort to further clarify what meta-analysis is and what it isn't, it is useful to compare and contrast the meta-analytic perspective with two more familiar entities: the traditional narrative literary review and the typical primary-level hypothesis test.

NARRATIVE REVIEWS VERSUS META-ANALYSIS

Although the differences between narrative reviews and meta-analytic reviews are of primary interest here, we can identify several interesting similarities. Both narrative research reviews and meta-analytic integrations operate on the implicit assumption that previous studies within a given research domain can illuminate facets of the phenomenon under study, and that the summary and integration of the results of these previous studies should serve to further our understanding of the phenomenon under study beyond the level of understanding achieved with any single investigation. Moreover, like any specific implementation of the scientific method, a review of research within a given domain is likely to involve a sequence of established phases or stages of inquiry (Rosenthal & Rosnow, 1985). For both narrative and meta-analytic reviews, a general sequence of steps can be identified that moves the (narrative or meta-analytic) reviewer from initial questions to final conclusions. For example, Cooper (1984) identified the progression from *problem formulation, data collection, data evaluation, analysis, and interpretation* to *public presentation* as characterizing both narrative and meta-analytic research reviews. Similarly, Light and Pillemer (1984) identified a set of issues progressing from *formulating a precise question, conceptual focus* (exploratory vs. hypothesis testing), *selecting studies,* and *generalizing results* to *relating study characteristics to outcomes.* What these and other frameworks suggest is that both narrative reviews and meta-analyses, as well as primary-level studies, can involve similar fundamental steps from defining the hypothesis of interest, to collecting the relevant previous research, and somehow integrating the results of this previous research into an illuminating summary of the research domain.

In spite of the fact that all reviews of a given research domain may start with the same basic goals, one critical problem is that all reviews of a given research domain might not end with the same basic conclusions. Light and Pillemer (1984) described the type of conflict that can emerge between two traditional narrative reviews of the same research literature. For example, Munsinger (1974) reported a narrative review of a sample of studies that examined the cognitive performance of adopted children, and concluded that environmental effects on children's IQ test scores were small. Kamin (1978) reported a narrative review of the same sample of studies, and was led to the opposite conclusion. Kamin's reflection on the disagreement between the two narrative reviews was particularly disconcerting:

> Perhaps the major point to be made is that readers interested in evaluating the evidence on heritability of IQ ought not to depend on published summaries. Those who wish to speak or to teach accurately about what is and is not known have no realistic alternative but to read the literature themselves. (p. 200)

This suggests that experts in the field may, at least in some instances, be unable to summarize past research in a meaningful and illuminating way, using the narrative perspective. Moreover, how is the scholar who is interested in this type of research literature expected to conceptually, intuitively integrate the results of past research in the most useful way, when the experts appear unable to do so?

Perhaps even more serious than the disagreement in conclusions derived from two different narrative reviews is the disagreement in conclusions derived from a narrative review and those derived from a meta-analysis of the same research domain. For example, Maccoby & Jacklin (1974) conducted a narrative review of the literature on sex differences in conformity. This traditional review led Maccoby and Jacklin to conclude that the results of previous research suggested little or no consistent gender differences in conformity. However, subsequent meta-analytic reviews of this literature (e.g., Cooper, 1979; Eagly, 1978; Eagly & Carli, 1981) indicate that the greater susceptability of women to conformity and influence may be of moderate to small magnitude, but it is highly significant. More recently, Marks and Miller (1987) provided a narrative review of social projection effects (i.e., the tendency to "project" our own thoughts, feelings, and behaviors onto others, and to perceive other people as similar to ourselves). This traditional review found support for both motivational mechanisms (e.g., we need to feel that we're normal) and cognitive mechanisms (e.g., we make systematic mistakes in processing information about the prevalence of our own behavior) for this phenomenon. However, Mullen and Hu (1988) conducted two meta-analytic reviews of this same research literature, and found systematic support for cognitive mechanisms of social projection but not for motivational mechanisms.

To supplement these anectdotal illustrations of the different conclusions that can be produced by the two approaches to research integration, Cooper and Rosenthal (1980) had 41 faculty and graduate students read and integrate the results of the same seven studies of gender differences in task persistance. The results of these seven studies, though somewhat variable, did, in fact, support the hypothesis of greater task persistance on the part of females. Half of these "reviewers" were instructed in the use of meta-analytic techniques, whereas half were told to follow whatever procedures they would normally follow in trying to integrate and summarize a research domain. The traditional reviewers were significantly more likely to conclude that there was no support for the hypothesis of gender difference, relative to the (generally more accurate) meta-analytic reviewers. This provides a somewhat more controlled, and particularly disturbing, demonstration of the potential insensitivity of traditional narrative reviewing methods.

Why should we be more inclined to accept the conclusions derived from a meta-analytic review than those from a narrative review? To begin, it should be recognized that meta-analysis is nothing more than a tool; and like any tool, if used improperly or carelessly, it will result in a faulty product. One can easily imagine instances where a carefully done narrative review might be more informative than a slipshod and sloppy meta-analysis. We are all familiar with the dog-eared quip from computer science: "garbage in—garbage out." Some critics have suggested that a meta-analysis may be an exercise in futility, insofar as the thoughtless amalgamation of unrelated studies will be less than informative. As Strube and Hartman (1983) pointed out, "No amount of precision or quantification in the actual techniques can spare the reviewer the thought and planning that are required of *any* review" (p. 17). This is undeniably true of a poorly conducted meta-analysis; however, it is just as undeniably true of a poorly conducted traditional narrative review (Wachter, 1988).

Still, in the examples described previously, it seems certain that the narrative reviewers and the meta-analytic reviewers were equally conscientious, deliberate, and judicious. When the difference between conclusions of a narrative review and a meta-analytic review cannot be discounted as being due to irresponsible or injudicious application of either of the two approaches, why should the reader be more compelled by the meta-analytic review than by the traditional narrative review?

The foregoing discussion highlights three broad differences between traditional reviews and meta-analytic reviews. These differences are central tenets of the scientific method, and they argue on behalf of meta-analytic approaches over traditional narrative approaches. These differences are precision, objectivity, and replicability.

Precision. A properly conducted meta-analysis is nothing if it is not precise. The specific strategies and techniques for taking the statistical results of independent studies, converting these results to a few common statistical metrics, and then combining, integrating, and testing these statistical results are wholly designed to be rigorously precise. Alternatively, a traditional narrative review by definition does not have any precise method or strategy for abstracting the evidence from each study. A narrative review is ill-equipped to take into account the interrelations between significance level, sample size, and effect size, in the manner of a carefully conducted meta-analysis. This is reflected in the level of precision of the conclusions derived from the two approaches to summarizing research. While attempting to avoid a pejorative caricature, it is unfortunately too often the case that a traditional narrative review ends with a statement that there does seem to be something to the hypothesis under consideration, but that there is some inconsistency in the results of past research, rounding off with a "clarion call" for more research. In contrast, as detailed throughout later chapters in this book, the meta-analytic summary and integration of a research domain provides precise gauges of the significance and the magnitude of the phenomenon under consideration, of the variability of these study outcomes, and, perhaps most importantly, of the extent to which these study outcomes can be accounted for in theoretically and practically illuminating ways.

Objectivity. The rules and standards for including studies in the review process, abstracting results from them, and weighing them in the final integration are never made explicit in a traditional narrative review. In contrast, these rules and standards must be made explicit in any meta-analytic review. The disagreement between narrative literary reviews, such as that illustrated by the debate between Munsinger (1974) and Kamin (1978), which was described previously, is due in large part to the inarticulate and subjective rules that guide the narrative reviewers in their separate (and sometimes irreconcilable) approaches to a research domain. The rules and strategies that are followed in a meta-analysis for including, abstracting, weighting, and integrating studies are generally more objective. Even where some subjective judgments may have to be made in conducting a meta-analysis, these subjective judgments are made public, articulated, and submitted to careful scrutiny by the reviewers and audience of the meta-analysis. For example, one might argue that the decisions as to which studies should be included in the review of a particular research domain is a subjective decision on the part of either type of reviewer. However, this decision is likely to remain implicit and inarticulate in the case of a traditional narrative review, whereas it must be made explicit and public in the case of a meta-analytic review.

Replicability. One product of the objective and public nature of the rules and strategies that guide a meta-analysis is a high degree of replicability. In other words, the conclusions derived from a meta-analytic review of a given research domain are the same conclusions at which anyone would have arrived if they had included the same studies and followed the same rules for integrating study outcomes. On the other hand, because the rules and strategies of the traditional narrative review are not objective and public, the same research domain can be reviewed by two different narrative reviewers and give rise to two very different conclusions.

Although it may seem tautological to emphasize how someone following the same rules will arrive at the same conclusions, it is important to recognize two implications of this issue. First, this is a critical difference between narrative reviews and meta-analytic reviews: The results of a meta-analytic review can be replicated, whereas the results of a traditional narrative review might not be replicated. Second, those researchers who wish to criticize a meta-analysis because of the particular applications of a particular rule or strategy will be able to determine the consequences of the particular application of the rule or strategy in question. That is, the critic will only have to identify the decision in question and conduct the alternative meta-analysis, using the application of the rule or strategy that they feel is more defensible for that set of studies. Thus, the source of any disagreement between meta-analytic summaries of a given research domain can be traced to the application of some objective and explicit rule or strategy (see Wanous, Sullivan, &

Malinak, 1989). Thereafter, the theoretical implications of that particular difference in application can be considered, and the general summary of the research domain can be adjusted to accommodate the two different sampling, coding, or analytic strategies.

Finally, it should be recognized that by acquiring the skills and the perspective of a meta-analyst, one does not lose the useful skills and perspectives of a good narrative reviewer. Our capacities to think critically, to generate heuristic explanations for unanticipated patterns, and to integrate specific results into a broader theoretical context are not impaired simply because we become more careful about how we integrate the statistical results of independent studies. The most informative summary of a research domain is going to be the one that combines the sound and responsible application of meta-analytic techniques with the careful and thoughtful approach that any reviewer takes to an interesting puzzle.

PRIMARY-LEVEL HYPOTHESIS TESTS VERSUS META-ANALYSIS

Although the similarities between primary-level hypothesis tests and meta-analysis are of primary interest here, we can identify a few interesting differences. The large and obvious difference between primary-level research and meta-analysis is the unit of analysis. To state the obvious, the unit of analysis in a single primary-level hypothesis test is the subject—the student, patient, or some other single observational entity; the primary level of analysis is typically geared toward the summary and integration of the responses of several separate individuals. Alternatively, the unit of analysis in the meta-analysis is the results of a study; the meta-analysis is typically geared toward the summary and integration of the results of several separate studies.

Another, less obvious difference between primary-level studies and meta-analysis is in the application of appropriate statistical techniques. Some meta-analysts (e.g., Glass, McGaw, & Smith, 1981) simply apply the statistical techniques developed for primary-level hypothesis tests to meta-analytic data points. However, there are some implications of the differences in units of analysis (described earlier) that create some problems in the application of primary-level inferential statistics to meta-analytic data points. For example, primary level inferential statistics assume homogeneity of variance for the data points included in the analysis. However, because the statistical tests of the hypothesis included in a meta-analysis will usually be based upon differing sample sizes, these statistical tests will be likely to have different sampling variances, thus violating the assumption of homogeneity of variance. It is for this type of reason that statistical procedures that were developed specifically for meta-analysis are argued to be more appropriate for the summary and integration of meta-analytic data points (cf. Cooper, 1984; Hedges & Olkin, 1985; R. Rosenthal, 1984; Strube & Hartman, 1983). As Hedges and Olkin (1985) observed, "There does not appear to be any rigorously defensible argument for the use of conventional t-tests, analyses of variance, or regression analyses to analyze effect sizes or correlations" (p. 12).

In spite of these procedural differences, there are several important conceptual parallels between the development of a successful primary-level hypothesis test and the development of a successful meta-analysis. Recall that a consistent sequence of events was identified earlier for narrative reviews and meta-analytic reviews, as well as for primary-level investigations. It bears repeating that both primary-level studies and meta-analytic integrations will involve the scholar in a similar sequence of tasks. These tasks include a careful definition of the hypothesis to be examined, the collection and assembling of the units of measurement, putting the units of measurement on a common metric, and asking three general analytic questions. These tasks are detailed hereafter.

Definition of Hypothesis. The successful primary-level study begins with a well-defined hypothesis. A well-defined hypothesis provides a clear specification of the direction in which (and, if possible, the extent to which) Y will vary as a function of X. A well-defined hypothesis also provides a clear specification of precisely what is meant by the variables of interest, X and Y. This means that a well-defined hypothesis test moves us beyond conceptual definitions (i.e., definitions of a variable or concept in terms of other concepts) into useful operational definitions (i.e., definitions of a variable or concept in terms of procedures or operations of measurement or manipulation). For example, we may start by conceptually defining *intelligence* as "how smart someone is", or "the ability to handle information." However, we cannot begin to engage in a well-defined primary-level hypothesis test until we operationally define *intelligence* in terms of something like "scores on the Wechsler Adult Intelligence Scale," or "grade point average." Any operational definition runs the risk of being incomplete or imperfect. But the hallmark of a well-defined hypothesis test (and a fruitful research paradigm that is comprised of well-defined hypothesis tests) is that it consistently and thoughtfully incorporates careful and defensible operationalizations of the variables of interest.

We often use the term *exploratory study* to refer to primary-level studies that do not clearly specify particular influence of some cause on some effect. Exploratory studies can be illuminating, and they can lead to new and interesting avenues of research. However, what is unique about well-defined hypothesis tests is that they have the capacity to lend confirmation or disconfirmation to a particular theoretical point of view. If the expected effect obtains, this suggests that the theory that guided the formulation of the hypothesis deserves our confidence; if the expected effect does not obtain, this suggests that the theory that guided the formulation of the hypothesis may not deserve our confidence. Perhaps the most informative type of primary-level hypothesis test is what Platt (1964) identified as a *strong inference test*. This describes the scenario where two different theoretical perspectives lead to two mutually exclusive hypotheses, and a single investigation is able to pit these two points of view against each other. With a strong inference test, whichever pattern obtains in the data will be informative, because it will support either one point of view or the other.

Just as with successful primary-level hypothesis tests, a successful meta-analysis must incorporate careful definition of the hypothesis to be examined. Fiske (1983) noted that meta-analysis requires researchers to deal with problems of fundamental categories and general definitions, and to exercise greater concern for the operationalization of independent variables and dependent variables. As indicated earlier, Smith and Glass's (1977) meta-analysis of psychotherapy outcome studies was a provocative impetus for the development and application of meta-analytic techniques; however, one of the criticisms that has been leveled against this creative effort is that it lumped too many different independent variables together (i.e., therapy approaches that were widely divergent), as well as too many dependent variables together (i.e., outcome measures that were conceptually unrelated; cf. Fiske, 1983; Strube & Hartman, 1983; Wilson & Rachman, 1983). This is what is sometimes referred to in an off-hand manner as "adding apples and oranges", indicating that the operationalizations employed by the studies included in a given meta-analysis may be undefensibly divergent. This point is well taken. Just as a successful primary-level hypothesis test must employ a clear and precise operationalization of the relevant variables, the meta-analysis that attempts to combine, summarize, and integrate the results of several such primary-level hypothesis tests is best advised to be consistent, and to focus on and include those primary-level hypothesis tests that have been clear and precise in their operationalizations.

Collecting Units of Measurement. Once clear operationalizations have been developed, and once a precise, specific hypothesis has been formulated, the primary-level investigator goes about the collecting and assembling of the units of measurement. This might amount to scheduling a large lecture hall, administering a questionnaire to a roomfull of college freshmen, and then collecting the completed questionnaires. Alternatively, this might amount to traveling to a primitive village in some remote region of the world and conducting extensive interviews with the inhabitants, which can be content analyzed later for references to family structure, ingroup–outgroup relations, and soon.

The process of collecting and assembling the units of measurement is critical for two reasons. First, and most obvious, is that without this step there will be no data to be analyzed. Second, and less obvious, is that the choices made at this step may influence the results obtained, and the generalizability of those results. Clearly, the references to family structure and ingroup–outgroup relations derived from extensive interview might have been very different if derived from transcripts of interviews with the college freshmen, and the questionnaire responses might have been very different if derived from the villagers. Another way of considering this issue is to imagine whether the interviews conducted with college freshmen would tell you much about the family structure and ingroup–outgroup relations of the villagers.

In a similar vein, the studies that are included in a meta-analytic data base exert the same types of influence. First, if there are no studies examining a particular hypothesis test, it is, by definition, impossible to perform a meta-analytic integration of such hypothesis tests. Second, the particular studies that are included may have implications for the generality of results obtained. For example, if one is interested in the effects of leadership behaviors on subordinate performance, one's results might be influenced by whether studies are conducted in Western, capitalist countries or communist block countries, due to differences in the two cultures in definitions and roles of leaders and subordinates. One's results might be influenced by whether studies were published in the mainstream scholarly journals (with, presumably, the highest editorial standards and the most stringent acceptance rate) or less prestigious, pay-for-publication journals (with, presumably, less demanding editorial standards and a more lenient acceptance rate). Thus, analogous to the implementation of primary-level hypothesis testing, the collection and assembling of the units of measurement in a meta-analysis may exert some influence on the results of the technique.

Common Metrics. In primary-level hypothesis tests, the units of measurement are often on a common metric at the time of data collection. For example, if you were measuring the height of children in different schools, the yardstick you used in each school was on a common metric of inches. However, if you find yourself in one school that has a meter stick but no yardsticks, you could measure the height of children in that school in centimeters and then convert these measures into inches, so that all of the numbers representing children's height would be on the same scale. Similarly, the primary-level investigator who wishes to gauge the effects of some intervention on the performance of a number of runners might express the runners' post-intervention running times in terms of a difference between pre- and post-intervention speeds.

The meta-analyst who wishes to integrate the results of three separate tests of the effects of *X* on *Y* is confronted with a similar problem when one of the primary-level hypothesis tests derived an *F* test to gauge this effect, the second test used a *t* test, and the third test reported a chi-square. There are various techniques, which we consider in detail in later chapters, for transforming these different statistics into a few common metrics, such as *Z*s and one-tailed probability levels for significance, and product–moment correlations, Fisher's *Z*s, and Cohen's *d*s for effect size.

Three Analytic Questions. Once the primary-level researcher has collected the units of measurement and placed them on a common metric, three general analytic questions can be considered. The first, *central tendency,* is a way of representing the typical response, and is typically gauged in terms of things like mean, median, and modal responses. The second, *variability,* is a way of representing the typical difference from the typical result, and is gauged in terms of indices like range and standard deviation. The third, *prediction,* is a way of trying to explain and account for the variability around the typical result. This is gauged in terms of correlations between the measures of interest and some other meaningful measures, or in terms of differences between means in the measurement for different subgroups of respondents.

Once again, a parallel between the primary level of analysis and the meta-analysis can be identified. For central tendency, the meta-analytic integration provides indices of the combined probability of finding results as extreme, or more extreme, as those obtained in the meta-analytic data base; in addition, the meta-analytic integration provides indices of the mean magnitude of the observed effects. For variability, the meta-analytic integration provides indices of the likelihood that the various included hypothesis tests were likely to have been sampled from different populations. That is, this indicates the extent to which the included studies are significantly heterogeneous. Finally, for prediction, the meta-analytic integration provides indices of the extent to which significance levels and effect sizes vary in meaningful, predictable ways as a function of practically important or theoretically illuminating variables.

The operation of each of these tasks at the meta-analytic level is the focus of separate chapters throughout the remainder of this book. Although the specific implementations that have been developed to handle meta-analytic problems are in some ways very different from the implementations developed at the primary level of analysis, the point is that the tasks and the problems confronting the meta-analyst are similar in many ways to those confronting the primary-level researcher. Despite the substantial procedural differences between primary-level research and meta-analysis, we often see a considerable similarity in the basic strategies of the two approaches. As Glass (1977) noted, "The accumulated findings of . . . studies should be regarded as complex data points, no more comprehensible without statistical analysis than hundreds of data points in a single study" (p. 352).

So, we begin with a picture of meta-analysis as a set of statistical procedures, and a general perspective, for the precise summary and integration of the results of previous research. On the one hand, this enterprise is similar to the traditional narrative literary review in some of its goals and assumptions, but it differs in its capacity to be more precise, objective, and replicable. On the other hand, this approach operates at a different level of analysis than the typical primary-level hypothesis test, and yet it shares a considerable degree of the strategy of the primary level study. In a sense, meta-analysis represents an application of the basic strategy of the primary-level hypothesis test to the goals of the traditional narrative review.

2

Defining the Hypothesis Test

It is a good plan in any study to observe for a while carefully but with no definite purpose.

—Charles Horton Cooley (1927)

Research without an actively selective point of view becomes the dirty bag of an idiot, filled with bits of pebbles, straws, feathers, and other random hoardings.

—Robert Lynd (1939)

Before the meta-analytic reviewer can begin transforming study statistics to some common metric, before relevant studies can be selected into or rejected from the meta-analytic data base, the hypothesis test of interest must be defined with care and precision. We cannot determine which study statistics, or which studies, are appropriate for inclusion into the meta-analysis unless we know precisely which hypothesis is being tested at the primary level.

One of the patent criticisms leveled against applications of meta-analytic techniques is the indictment of "adding apples and oranges." That is, in an effort to pull together as many related studies as possible, some meta-analytic integrations have lumped into one analysis studies that were vastly different in operationalizations, procedures, and theoretical focus. As we will see, there are good reasons to try to be as inclusive as possible. At the same time, however, the present chapter emphasizes the importance of maintaining a conceptual focus on well-defined hypotheses as a prequisite for an illuminating and defensible meta-analysis. In a recent policy statement, the editors of *Psychological Bulletin* (Masters, Cantor, & Hedges, 1988) highlighted the importance of careful conceptualization:

> The use of statistical methodology in accomplishing a literature review must be driven by conceptualization and judgment. . . . A review with a careful meta-analysis but weak or irrelevant conceptualization offers little for the advance or the consolidation of knowledge. (p. 4)

There are four specific facets of primary-level research that deserve special attention in defining the hypothesis test: the dependent variable, the independent variable, sampling and procedure, and the statistical test. We consider each of these elements in turn.

DEPENDENT VARIABLE

One could make the case that careful definition of the dependent variable examined in the primary-level studies is the most critical step in the initial stages of a meta-analysis. The dependent variable provides the gauge of the phenomenon that the reviewer is trying to integrate and understand. The dependent variable is generally conceptualized as the effect that results from some specific cause. We often begin with some form of *conceptual definition* of the behavior of interest. That is, we start off with a general notion of the phenomenon to be scrutinized, a definition of the concept in terms of other concepts. However, in order to move to the data collection stage, the primary-level researcher must develop and implement an *operational definition* of the behavior of interest. That is, the concept is defined in terms of procedures or operations of measurement. Thus, the primary-level researcher may have an interest in studying the effects of the number of toys a child has on how intelligent he or she becomes. This notion of "intelligence" may begin as a general notion of being able to handle information effectively (conceptual definition). However, before the researcher can move into the data collection phase, this conceptual definition would have to be transformed into measured scores on the Wechsler Intelligence Scale for Children, the Wechsler Preschool and Primary Scale of Intelligence, or some other standardized measurement procedure (operational definition).

The meta-analytic reviewer is in a rather unusual position with regard to operationalizations of the dependent variable. On the one hand, the reviewer has the luxury of being able to specify any one of several available operationalizations of most dependent variables. If one particular operationalization of the dependent variable has received severe criticism on methodological or theoretical grounds, the reviewer might well be advised to direct their attention elsewhere if their interest is in dealing with clear and unequivocal indices of the behavior in question. Of course, if one's interest lies in the controversy surrounding a problematic operationalization, then one is perfectly able to focus one's meta-analysis on those studies that employ the problematic measure; but the opportunity to make this choice highlights the value of developing a clear focus in these early stages of the meta-analysis. The goals of sorting out and clarifying the controversies around a problematic and poorly-understood measure are very different from the goals of trying to integrate the significance, magnitude, and variability of effects found in an accepted and well-understood measure. Both goals are legitimate, and the pursuit of both goals could be fruitful; but, even if two such meta-analytic integrations were to be published in the same paper, they would constitute two separate and distinct integrations.

On the other hand, the meta-analyst is typically stuck with both the strengths and the liabilities of the dependent measures that are available in the literature. If a given research literature is based on a single, clear, and defensible operationalization of the behavior of interest, then the task of selecting the operationalization of the dependent variable is already completed before it started. However, if a given research literature is based on several conceptually related but distinct operationalizations, the task of selecting and defending the operationalization of the dependent variable may be difficult, at best. Indeed, one will often find, unfortunately, that the most compelling and defensible operationalization of the dependent variable is used least often (perhaps because it requires access to special or expensive equipment or training). In such an instance, the number of studies that have measured and tested the hypothesis of real interest may be so few that a meta-analysis conducted on this small and new research paradigm may be (perhaps inappropriately) criticized as "premature."

It should be recognized that the meta-analytic techniques themselves (like any statistical formulae) do not care how many data points there are. As R. Rosenthal (1984) has clearly shown, one can easily meta-analytically integrate the results of only two hypothesis tests. Meta-analyses conducted on small research literatures can, in fact, be both practically and theoretically informative, especially when it can be demonstrated that an exhaustive search reveals that the universe of includable studies really is quite small (e.g., Mullen, Copper, & Driskell, 1989; Driskell & Mullen, 1989). Nonetheless, it is undeniable that a meta-analysis conducted on 500 hypothesis tests is likely to produce results that are more compelling than one conducted on 5 hypothesis tests. Thus, efforts to focus on clear and consistent operationalization of the dependent variable provide something of a double-edged sword: They make for a tighter, more defensible meta-analysis, but they are likely to reduce the meta-analytic data base.

INDEPENDENT VARIABLE

Careful definition of the independent variable in a meta-analysis is probably almost, if not just, as important as definition of the dependent variable. The independent variable is generally conceived of as the cause that produces a particular effect. Precision in the treatment of both dependent and independent variables is critically important. In the previous section we considered how the primary-level researcher begins with a general conceptual notion of the variable of interest (*conceptual definition*). In the case of independent variables, we start off with a general notion of the situational variation that causes something else to occur, a definition of the concept in terms of other concepts. However, once again, in order to move to the data-collection stage, the primary-level researcher must develop and implement an *operational definition* of the situational variation. That is, the concept is defined in terms of procedures or operations of measurement or manipulation. Recall the primary-level researcher's interest in the effects of a child's number of toys on how intelligent he or she becomes. This notion of "number of toys" may begin as a general notion of having lots of play resources (conceptual definition). However, before the researcher can move into the data collection phase, this conceptual definition would give rise to a tabulation of the number of toy dolls and toy cars (a measurement operational definition), or the controlled introduction or nonintroduction of children's computers (a manipulation operational definition).

Once again, the meta-analyst has the luxury to select the most theoretically relevant and illuminating operationalizations, while having to live with the consequences of small research literatures or problematic manipulations.

SAMPLING AND PROCEDURE

A third area that deserves special attention in defining the hypothesis test is the general area of sampling and procedure. By sampling, we refer to the process by which the units of analysis are included in the study. In other words, the subjects may be recruited through the standard university undergraduate Introductory Psychology human subject pool, or they may be patients with a particular set of symptoms, selected from a particular type of hospital. By procedure, we refer to the mechanical, sequential details of the primary-level investigation. Several aspects of the procedure contribute to the operationalization of the dependent and independent variables: Recall that the typical description of an operational definition is that it is a definition of a term or variable in terms of some *procedure* or operation of measurement or manipulation. However, there are several things about study procedures that do not directly determine the operationalizations of dependent and independent variables, yet do serve to define and delineate the hypothesis test as a whole.

With regard to sampling, the hypothesis test being examined at the primary level may vary substantially from one study to the next if the subject samples are terribly different. For example, the primary-level researcher may be interested in testing a particular hypothesis about the strategies of information integration in a complex judgment task. Studies that have attempted to examine this issue with samples of very young children may actually be testing a very different hypothesis from studies that have attempted to examine this issue with samples of college students. In spite of identical operationalizations of the dependent and independent variables, the nature of the task and the dimensions of response may be functionally very different to children who do not have the same cognitive sophistication as the college students.

For example, Rosenthal and Rosnow (1975) have examined in depth the differences between volunteer subjects and nonvolunteer subjects. Generally, volunteer subjects may be more responsive to task-orienting cues or demand characteristics. There is also the distinct possibility that subjects who volunteer themselves into a particular procedure are more interested in, or somehow predispositionally sensitive to, the phenomenon being suggested to them in the experiment's title or description. Alternatively, subjects may be likely to differ in their responses to the experimental setting as a function of some simple basic demographic features, like gender or race. In fact, some meta-analysts have put these possible differences to good use by incorporating these attributes into meta-analytic integrations (e.g., Eagly, 1986; Miller & Davidson-

Podgorny, 1987). We return later to a discussion of the techniques by which these sampling differences can be examined and tested at the meta-analytic level. However, the point in the present context is that we should be very careful about the potential influence of sampling differences on the meta-analytic enterprise. Depending on the phenomenon under consideration, the meta-analytic reviewer might be well advised to restrict their definition of the hypothesis test to those primary-level studies that employed particular sampling procedures.

The same may apply for other details of procedure. Once again, the influence of procedure on the test of a hypothesis may go beyond the operationalization of the dependent and independent variables. For example, the same measurements and manipulations might be implemented individually or in the context of groups. While the presence and the size of a group may be another study attribute that can help to illuminate underlying processes (cf. Mullen, Johnson, & Salas, 1989; Mullen, Salas, & Driskell, 1989), in some contexts it may make more sense to restrict the focus of the meta-analysis to those studies conducted in individual one-on-one sessions.

Another procedural detail that can exert substantial effects is the structure of the research design. For example, the same measurements and manipulations could be conducted in either a within-subjects design (in which multiple measurements are taken from each subject after exposure to each level of the manipulated independent variable) or a between-subjects design (in which each subject is only exposed to one level of the manipulated independent variable, followed by only one measurement of the dependent variable). Within-subjects designs are more powerful, in the sense that variability between levels of the independent variable cannot be attributed to individual differences (because each subject, with all of their individual differences, was exposed to all levels of the independent variable). However, within-subjects designs have been demonstrated to "sensitize" subjects so that they come to respond more to the treatment than if they had not been previously exposed to the stimulus materials (Rosenthal & Rosnow, 1984; Rosnow & Suls, 1970). Campbell and Stanley (1963) described the within-subjects, repeated measures design as being sensitive to multiple treatment interference, as well as the possibility of the reaction or interaction effect of testing. By way of illustration, Mullen's (1985) integration of the effects of the strength and immediacy of sources of social impact in group contexts suggested that strength and immediacy seemed to exert their strongest effects in the context of within-subjects designs. Thus, in defining the hypothesis of interest, it may be critical to specify additional facets of the procedure.

ISSUES RELEVANT TO OPERATIONALIZATION AND PROCEDURE

One issue that must be addressed is that some research literatures do not revolve around the cause–effect conceptualization of independent variable and dependent variable. Some research literatures deal exclusively with the task of determining the significance and the magnitude of the relation between two sets of measures, with no assumptions that the first measure is some kind of independent variable that causes the second, dependent variable. However, even in such research literatures, the necessity of careful attention to the operationalizations of the two variables under consideration is critical. An example of this type of research literature, which has yielded to meta-analytic scrutiny in recent years, is the domain of validity generalization in industrial/organizational psychology. For example, a series of primary-level studies may have examined the extent to which scores on a particular selection test accurately predict later performance of job applicants in the work setting. The selection test is not a genuine independent variable, and the subsequent job performance is not a genuine dependent variable. Nonetheless, successful meta-analytic integrations of this type of research literature (e.g., Hunter, Schmidt, & Hunter, 1979; Schmidt, Hunter, & Pearlman, 1981) have had to deal with the fundamental problem of operationalizations of the two sets of measures as delineated previously. A similar example of the necessity for attention to details of operational definitions in a nonexperimental research domain is found in Lord, deVader, and Alliger's (1986) integration of research on personality traits and leadership perceptions.

It is illustrative to consider the consequences of an excessively wide focus in operational definitions in a meta-analytic integrations. As mentioned earlier, the meta-analysis of psychotherapy outcome by Smith and Glass (1977) was crucial in fostering widespread attention to, and consideration of, the techniques and strategies that we have come to identify as *meta analysis*. Moreover, it should be noted that many of the general conclusions from this important effort have since been replicated in more narrowly focused meta-analyses (e.g., Andrews & Harvey, 1981; Shapiro & Shapiro, 1983). However, many of the criticisms of the original Smith and Glass integration revolve around this effort's terribly inclusive, almost unselective, definition of the hypothesis being tested. Smith and Glass were very clear on this point: Their goal was to provide a general, global integration of the effects of psychotherapy, across treatment modalities and across symptom typologies. However, again and again, critics of this early effort emphasized that the differences in "treatment" across different theoretical and practical perspectives, and the differences in "psychotherapy outcome" across different symptom typologies and measurement procedures, made the aggregation of these diverse outcomes untenable (cf. Cook & Leviton, 1980; Eysenck, 1978; Rachman & Wilson, 1980). It was this lack of focus in these early stages of defining the hypothesis, and the deluge of critical responses on this point, that lends support to the caricature of meta-analysis as being a thoughtless adding of apples and oranges. As Mintz (1983) noted, "Even if the computer is willing to compare aspirin and bypass surgery with the type of disorder "partialled out," it makes no sense to do so" (p. 74).

At the opposite end of the scale is the meta-analytic integration that accepts the liabilities of a small data base but goes on to derive some interesting and illuminating conclusions from a research literature by virtue of the careful focus on operationalization. For example, Mullen & Felleman (in press) examined the effects of "tripling" in college dormitories (i.e., assigning three people to live in a dormitory room designed for two people). These analyses revealed that the aversive consequences of tripling were significant and of moderate magnitude, for each of four distinct operationalizations of "responses to tripling". Moreover, this meta-analytic integration revealed, for the first time, the highly significant moderation of the effects of tripling by the length of the dormitory corridor. Whereas previous primary-level research (e.g., Baum, Aiello, & Calesnick, 1978) had demonstrated more crowding effects as a result of long, oppressive corridors, the Mullen and Felleman (in press) meta-analysis was the first demonstration of the exacerbation of the effects of tripling by corridor length. The considerable significance and magnitude of these effects have substantive theoretical and practical implications that are not mitigated by the fact that this small research domain only comprised 10 studies.

It should be noted that some meta-analysts recommend the widest, most inclusive treatment of operational definitions allowable by the research domain under consideration. For example, Cooper (1984) provides a compelling argument that *multiple operationism* strengthens a meta-analysis. Webb, Campbell, Schwartz, Sechrest, and Grove (1981) defined multiple operationism as the use of many measures that are supposed to gauge the same theoretical construct, but that have different patterns of irrelevant components (i.e., have different error variance). If one focuses on a particular operational definition of, for example, a dependent variable, then one's conclusions will be based on, and will cover up, the consistent influence of one particular pattern of unwanted, error variance. However, if one uses multiple operational definitions of the dependent variable, then one's results will not be contaminated by one particular pattern of error variance. In some instances, this might allow the theorist to converge or "triangulate" on the construct of interest. The obvious difficulty is that incorporating multiple sources or patterns of error variance may serve to more completely obscure the genuine patterns of the phenomenon, which may be sufficiently obscured already by a single pattern of error variance.

Let us examine the possible consequences of a lack of attention to the types of details specified earlier. Consider Fig. 2.1. The boxes in Fig. 2.1 represent separate tests of the hypothesis that *X* influences *Y*. Study outcomes are represented in this simple figure on a metric that arbitrarily ranges from 0 to 1: As we move up the vertical axis, we represent more of an effect of *X* on *Y;* as we move down the vertical axis, we represent less of an effect on *Y*. The horizontal axis represents some theoretically interesting moderator (e.g., the magnitude of the manipulation of *X;* the mean age of subjects in the study; etc.).

Two facets of this research domain become apparent from Fig. 2.1. First, the average effect of *X* on *Y* (in the arbitrary units depicted on the vertical axis) is about .50. Second, there is no apparent variation between study outcome (vertical axis) and moderator variable (horizontal axis), $r(10) = .190$. However, what if this set of 12 studies is really comprised of two distinct sets of six studies, each set different from the other in

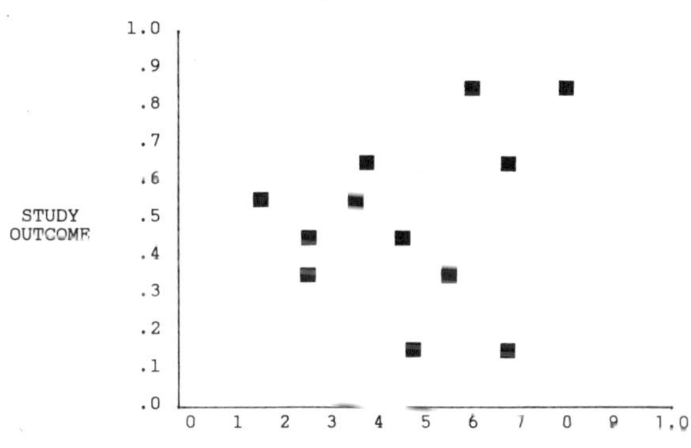

FIG. 2.1 Study outcomes and a moderator.

operationalization of the dependent variable, of the independent variable, or some important facet of sampling or procedure? Consider the same data, cast in Fig. 2.2, with the studies in one set represented by black circles, and the studies in the other set represented by white circles.

A very different picture emerges if we know that two distinct sets of hypothesis tests are included in this single aggregation. For example, the average effect of X on Y is .35 for the six white circle studies, whereas it is .65 for the six black circle studies. Moreover, study outcome seems to be positively related to the moderator for the black circle studies, $r(4) = .717$, whereas it is negatively related to the moderator for the white circle studies, $r(4) = -.717$.

Certainly, if one were interested in just these types of differences between effects produced using different operationalizations, sampling, or procedures, one could in fact structure the meta-analysis to incorporate both types of hypothesis tests and test for these differences. However, this proves the point about careful attention to the definition of the hypothesis test, because one would not be able to formulate and pursue such a (potentially illuminating) strategy at the meta-analytic level without having first done the careful and thoughtful labor of defining the relevant facets of the potentially includable hypothesis tests. Moreover, it is, unfortunately, too often the case that the meta-analyst who is trying to follow the "multiple operationism" dictum may simply not pursue the possibility of substantive differences between alternative operationalizations, sampling, or procedures, in the belief that this broad conceptual base only adds strength and power to the meta-analytic enterprise.

The best strategy, incorporating Cooper's (1984) emphasis on multiple operationism and the present author's emphasis on theoretical focus, is to "cast the widest net possible," and begin by defining the hypothesis in the broadest, or most general, way. However, differences between

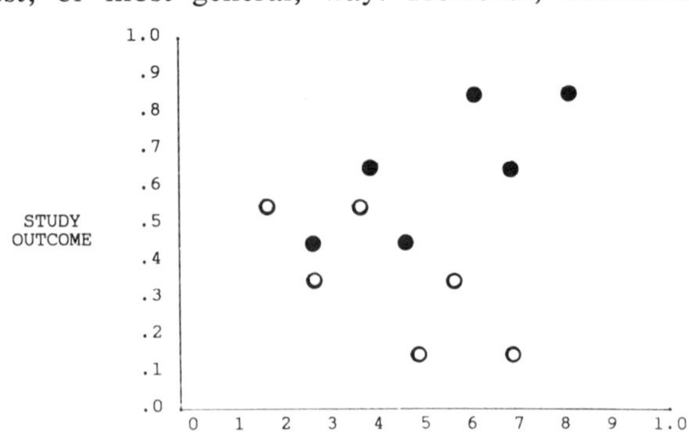

FIG. 2.2 Study outcomes and a moderator, with two sets of studies.

always less than .50 when results are in the "expected direction" (i.e., when results support the hypothesis under consideration); however, one-tailed p values are always greater than .50 when results are in the "unexpected direction" (i.e., when results are inconsistent with the hypothesis under consideration). Reported two-tailed p values supporting the hypothesis under consideration should be halved; reported two-tailed p values inconsistent with the hypothesis under consideration should be halved and then subtracted from 1.00. As suggested by Overall and Rhoades (1986), one should be careful, in applying this adjustment of two-tailed p values, that the p values with which one starts are truly two-tailed; following this halving procedure on a p value that is already one-tailed results in a dangerously unconservative (and probably meaningless) "half-tailed" hypothesis test statistic.

However, certain restrictions apply to the inclusion of F tests and chi-squares. The only F tests that are appropriate for inclusion in meta-analyses are Fs with 1 degree of freedom in the numerator; an F with more than 1 degree of freedom in the numerator may be a useful statistic in some contexts, but it often represents an unfocussed, general comparison between several means. For example, Table 2.1 presents the results of analyses of variance for four separate studies.

Suppose that the hypothesis of interest is that an increase in X will lead to an increase in Y. Note that all four studies test the effects of X on Y with a one-way, three-condition analysis of variance design. Also note that each study's analysis yields an $F(2,15) = 7.5$, $p = .0028$. Recall that degrees of freedom in the numerator for an F test are determined by (number of groups − 1), which in these instances yields $(3-1) = 2$. Degrees of freedom in the denominator are determined by (sample size − number of groups), which in these instances yields $(18-3) = 15$. However, these Fs with more than one degree of freedom in the numerator only tell us that the three cell means in a given study are

Table 2.1
Results of ANOVA for Four Separate Studies

	Study 1			*Study 2*			*Study 3*			*Study 4*		
n	6	6	6	6	6	6	6	6	6	6	6	6
Y	5	6	7	5	6	7	5	6	7	5	6	7
Value of X	1	2	3	3	2	1	None	1	2	1	2	10
MS(err)		0.80			0.80			0.80			0.80	
$F(2,15)$		7.50			7.50			7.50			7.50	
p		.0028			.0028			.0028			.0028	

significantly different from one another. They do not tell us anything about the particular pattern of these differences, or whether a given pattern of difference confirms or disconfirms the hypothesis that X causes an increase in Y.

For example, Study 1 shows that as the value of X (e.g., the strength of the independent variable manipulation) increases from 1 to 2 to 3, there is a corresponding increase in Y from 5 to 6 to 7. This linear trend can be tested with a more focused F test with only one degree of freedom in the numerator, which weights the cell means by orthogonal polynomial contrast weights (cf. Kirk, 1982, for a discussion of a priori linear contrasts). The corresponding F for linear contrast in Study 1 is $F(1,15) = 15.00$, $p = .0076$. Contrast this with the results for Study 2. The increase in X from 1 to 2 to 3 is associated with a decrease in Y from 7 to 6 to 5. The corresponding F for linear contrast in Study 2 is $F(1,15) = -15.00$, $p = .9924$. The results for Study 3 represent two functional manipulations of X, with values of X at 1 and 2, and a control condition with no manipulation of X. The difference between the Y produced where $X = 1$ and that where $X = 2$ is represented by an $F(1,15) = 3.75$, $p = .0360$. Finally, it is important to remember that orthogonal polynomial contrasts are designed to capture different intervals between conditions (as well as different cell sizes, although that doesn't enter in to this example; cf. Kirk, 1982,k Appendix C). Thus, for Study 4, the linear contrast for increases in X from 1 to 2 to 10 produces an F for linear contrast of $F(1,15) = 12.48$, $p = .0015$.

The point of this illustration is that the identical Fs with more than one degree of freedom in the numerator (sp., $F(2,15) = 7.50$) can represent very different patterns of results for different analyses [sp., $F(1,15) = 15.00$, $F(1,15) = -15.00$, $F(1,15) = 3.75$, and $F(1,15) = 12.48$]. It is for this reason that only Fs with one degree of freedom in the numerator can be included in meta-analyses as statistical tests of well-defined hypotheses.

A similar problem occurs with chi-squares that are based on more than one degree of freedom. Consider the two frequency tables presented in Table 2.2.

Both studies represented in this table conducted an opinion survey, where people belonging to one of two groups, or not belonging to either group, expressed an opinion agreeing with, disagreeing with, or uncertain about the issue being discussed. Suppose the hypothesis of interest is that people belonging to Group A are more likely to agree than disagree, compared to people belonging to Group B. Both frequency tables produce an identical, significant χ^2 (4) $= 14.84$, $p = .0073$. Recall that

Table 2.2
Frequency Tables for Two Studies

	Study 1				Study 2		
	Yes	*Unsure*	*No*		*Yes*	*Unsure*	*No*
Group A	3	4	20	A	3	20	4
No Group	5	6	24	No Group	5	24	6
Group B	7	8	4	B	7	4	8
χ^2		14.84				14.84	
p		.0073				.0073	

the degrees of freedom for a chi-square is defined as (number of rows − 1) × (number of columns − 1). In this instance, this yields $(3-1) \times (3-1) = 4$. However, analogous to the *F* test with more than one degree of freedom in the numerator, this significant chi-square with more than one degree of freedom only tells us that the frequency of agreeing, being uncertain, or disagreeing differs across the three types of subjects. These chi-squares do not tell us anything about the particular pattern of these differences, or whether a given pattern confirms or disconfirms the hypothesis that members or Group A are more likely to agree than disagree, compared to people belonging to Group B. Rosenthal and Rosnow (1984) provided a very useful discussion of the techniques and strategies for "breaking down" larger frequency tables into well-defined tests of specific hypotheses (the reader is referred to their presentation for a more complete discussion of these strategies). In the present context, the table describing the results of Study 1 can be broken down into a 2 × 2 table comparing the agreeing and disagreeing of subjects in Group A and Group B (i.e., the four corner cells in these particular tables). This produces a chi-square on one degree of freedom of $\chi^2 (1) = 9.174$, $p = .0012$, indicating that members of Group A really were more likely to agree than members of Group B. On the other hand, the frequency table for Study 2, represented by the same general chi-square test, produces a specific 2 × 2 hypothesis test of $\chi^2 (1) = -0.028$, $p = .5664$, indicating that members of Group A were slightly, nonsignificantly less likely to agree than members of Group B.

The point of this illustration is that identical chi-squares with more than one degree of freedom [sp., $\chi^2 (4) = 14.84$] can represent very different patterns of results for different analyses [sp., $\chi^2 (1) = 9.174$, and $\chi^2 (1) = -0.028$]. It is for this reason that only chi-squares with one degree of freedom can be included in the meta-analysis as specific tests of a well-defined hypothesis.

3

Retrieving the Studies

Attempt the end, and never stand to doubt,
Nothing's so hard but search will find it out.

—Robert Herrick (1648)

Through the process of defining the hypothesis test that will be integrated in the meta-analytic review, the reviewer has specified which studies will be appropriate for inclusion in the meta-analytic integration. Once we know precisely which hypothesis tests are of interest, we can begin the next step of identifying, locating, and procuring the studies that will be included in the meta-analysis. This involves much more than simply going out and photocopying the articles from the stacks in the library. As Rachman and Wilson (1980) have suggested, the results of a meta-analysis can be useless—and even counterproductive—if the meta-analysis is conducted upon an incomplete and biased data base.

There are several distinct facets to the problem of retrieving studies. These can be categorized into the processes of identifying the sources of studies, locating the studies, and dealing with the "file drawer problem."

IDENTIFYING SOURCES OF STUDIES

The user has to identify the sources of studies that will be included in the meta-analysis. Studies can be obtained from a variety of (overlapping) categories: articles published in professional journals, books; theses and dissertations; papers presented at regional and national conventions; archives; and, unpublished sources (unpublished manuscripts; manuscripts under editorial review). The reviewer must be particularly careful to allow a given hypothesis test to enter into the meta-analytic data base only once. For example, a specific hypothesis test of the effect of *X* on *Y* may be simultaneously represented in a researcher's PhD dissertation, a regional convention presentation, and a manuscript under editorial review. An exhaustive search could retrieve this particular hypothesis test from all three sources, and the reviewer would have to insure that the hypothesis test is represented only once in the meta-analytic database.

Some meta-analyses (e.g., Cooper, 1979) rely exclusively on studies obtained from published sources. This reliance on published sources may be based on the assumed higher quality of published research: Because these studies were (generally) accepted through the rigorous review of a group of academic peers, these published studies have apparently "passed the muster" and must be of acceptable methodological and theoretical quality. In addition, this reliance on published sources certainly limits one's data base to studies that are most easily available. Published sources can generally be obtained with the minimum of legwork, by simply walking through one's own university library or by working through one's university library interlibrary loan department.

However, some evidence indicates that publication policies may be biased in favor of reporting significant findings. (e.g., Atkinson, Furlong, & Wampold, 1982; Bakan, 1967; Chase & Chase, 1976; Fagley & McKinney, 1983; Greenwald, 1975; Rosenthal, 1966; Smart, 1964; Smith, 1980; Sterling, 1959; White 1982). For example, Greenwald (1975) conducted a survey of authors and reviewers for the *Journal of Personality and Social Psychology* during a 3-month period of 1973. The results of this survey indicated that researchers were considerably more likely to submit for publication studies that produced significant results than studies that failed to produce significant results. White's (1982) review of the effects of socioeconomic status on achievement revealed considerably stronger effects for hypothesis tests reported in books, moderate effects for hypothesis tests reported in journals, and considerably weaker effects in unpublished sources. Smith (1980) examined 10 research reviews where the size of the effect for published and unpublished hypothesis tests could be compared; in each case, the effects in published sources were stronger than those in unpublished theses and dissertations. R. Rosenthal (1984) reported the most comprehensive analysis to date of the differences between study outcomes from journals, books, theses, and unpublished sources. The results of his analyses suggest that journals, books, and unpublished studies may be largely indistinguishable from one another in study outcomes, whereas theses did tend to produce noticeably weaker effects.

These patterns suggest that a strict reliance on some sources to the exclusion of other sources could influence the estimates of the overall levels of significance and/or magnitudes of effect for the hypothesis test under consideration. Utilizing the full range of sources of studies will produce a more comprehensive and more accurate meta-analysis.

In spite of careful definition of the hypothesis test under consideration (as discussed at length in the previous chapter), the considerable research attention devoted to some topics may present the reviewer with the mixed blessing of an extremely large number of studies. One approach to this situation would be to simply conduct a massively large meta-analytic integration, including all of the retrieved studies. Another approach to this situation is to select (as best as possible) a random, representative sample from this population of studies, and conduct the meta-analysis on this smaller sample. This approach is illustrated in Rosenthal and Rubin's (1978) integrations of the research literatures on various types of interpersonal expectancy effects. Finally, in an approach that is most consistent with the last chapter's emphasis on detail and precision in definition, one might be best advised to sharpen one's conceptual focus and to perform separate meta-analyses on smaller bodies of more closely related research. The assumption underlying this approach would be that one's conceptualizations in the early stages of defining the hypothesis tests may not have been optimally focused and specific, and that it may be theoretically more illuminating to scrutinize a smaller and more comparable set of studies. This approach is illustrated in Hall's (1978, 1980) integrations of gender differences in the decoding of nonverbal cues. Whichever approach the reviewer takes to the troublesome luxury of a large number of studies, the reviewer must make his or her decision explicit, and justify it.

LOCATING STUDIES

Once the sources of studies have been identified, the legwork of actually obtaining copies of the studies begins. This stage of a meta-analysis can be tedious and time-consuming, and it requires the combined skills of a psycholinguist, an archeologist, and a private detective. A number of distinct strategies have been suggested (cf. Cooper, 1982; Glass et al.,

1981; R. Rosenthal, 1984), each of which should be implemented in the conduct of a full-scale and defensible meta-analysis.

Ancestry Approach. This strategy for locating studies utilizes the bibliographies and reference sections of relevant studies that have already been retrieved to locate earlier relevant studies. For example, suppose the reviewer is stimulated to attempt an integration of a particular research literature by an article published this year. The ancestry approach would lead the reviewer to locate all of the relevant studies cited in that recent paper. In turn, the reference sections from all of these earlier studies would be used to locate additional earlier relevant studies, and so on. Eventually, the ancestry approach will take the diligent reviewer back to the earliest recognized investigations of the phenomenon under consideration.

Descendancy Approach. The complement to the previous strategy, this approach utilizes indexing sources to retrieve subsequent relevant studies that have cited earlier relevant studies. For example, suppose the reviewer is stimulated to attempt an integration of a particular research literature by a classic article in the field, published 20 years ago. Indexing sources, such as the *Social Sciences Citation Index,* list the publications during each subsequent year that have cited that earlier target study. This provides the reviewer with an invaluable lead on more current work that has followed up on the seminal studies in a particular research paradigm. In turn, the relevant studies resulting from this descendency approach can themselves be subjected to a descendency search, to identify subsequent publications that have cited these follow-up studies. Eventually, the descendency approach will take the diligent reviewer forward to the most recent investigations of the phenomenon under consideration.

Abstracting Services. This approach allows the reviewer to seek out studies associated with key words and phrases. Abstracting services, such as *Psychological Abstracts, Sociological Abstracts, Dissertation Abstracts International, and Educational Resources Information Center (ERIC)*, reproduce the abstracts, or brief summaries, of most published research, and some unpublished research, in a given domain. In order to use one of these abstracting services, the reviewer must enter into the literature with a few key words or phrases. Working through the service's thesaurus (e.g., *Psychological Abstracts' Thesaurus of of Psychological Terms; Thesaurus of ERIC Descriptors*), the reviewer can usually develop a list of relevant key words. Then, the abstracts of studies published within a given year that have been classified in terms of these key words can be read with an eye toward making a preliminary determination of whether the study seems like it might be relevant to the hypothesis test being integrated in the meta-analysis. Eventually, this approach will uncover all of the studies that have been classified under the critical key words.

On-Line Computer Searches. Although sources like *Social Sciences Citation Index* and *Psychological Abstracts* can be used "by hand," active meta-analysts are also finding it increasingly easier to allow information scientists to perform these types of searches through on-line computer systems (cf. Atkinson & Dolan, 1983; M. Rosenthal, 1984). These on-line searching services perform functionally the same task that one would perform by hand for descendency searches and abstracting services searches, but at a tremendous savings of time and energy.

The "Invisible College." Price (1966) identified the invisible college as the informal network of scientists working on the same problem. Budding meta-analysts are often pleasantly surprised at how fruitful a courteous request to the more active researchers in a given area can be in accessing new and/or unpublished studies. Letters, phone calls, and conversations at professional conferences may be likely to uncover relevant hypothesis tests and manuscripts at various stages of being "in the works."

Browsing. There is a lot to be said for the lost art of browsing. Browsing, especially through the publications of peripherally related disciplines, is a terribly inefficient means of locating studies when compared to the speed and power of the on-line computer search. However, particularly in literatures that deal with applied, naturalistic, or interdisciplinary phenomena, there may be a pocket of research testing the same hypothesis within a different research domain that neither cites nor is cited by the research literature within which the reviewer is working. In such instances, unfortunately, there is no way to uncover these relevant and includable studies except through browsing.

In addition, proceedings from professional association conventions may provide access to otherwise unpublished research. However, convention proceedings are often published with little or no topical subject indexing, which makes browsing the only way to determine what was presented. In addition, most of us have probably had the experience of having a convention presentation scheduled in a session of papers that were less than meaningfully related to one another. In such instances, the title of the session may be less than perfectly informative regarding the nature of the papers included in that session. Again, this argues on behalf of browsing as the only strategy to uncover otherwise inaccessible studies.

Problems with the Various Strategies. Each of the foregoing strategies has the potential to uncover relevant, includable studies that might not have been located in any of the other approaches, and for that reason each of the foregoing strategies is a good one. However, there are difficulties and problems involved in each of these approaches that should be highlighted. For example, the ancestry and descedency approaches share the potential difficulty of yielding predominantly published articles. If one's concern is to incorporate unpublished papers and theses, then one may have to complement these strategies with the use of the *Dissertation Abstracts International* abstracting service, browsing through convention proceedings, and querying active researchers in the area.

In addition, the ancestry and descendency approaches share the problem of irrelevant citations. Many citations in scholarly journals may be useful or important in establishing some broader theoretical context, or defending a particular methodological or statistical decision for the primary-level study. However, these citations may be "irrelevant" to the meta-analyst in the sense that these cited studies are not additional relevant tests of the hypothesis under consideration. In the case of the ancestry strategy, this means that the reviewer may devote a lot of time to tracking down earlier studies that provide a context, and serve as an impetus, for the hypothesis under consideration, but that do not test this hypothesis directly. In the case of the descendency strategy, this means that the reviewer may devote a lot of time to tracking down subsequent studies that refer to the interesting hypothesis, but that do not test the hypothesis directly.

Abstracting services can be an invaluable aid to locating relevant studies—as long as you start with the right key words. Sometimes, this simply requires becomming familiar with the classifications, categorizations, and labels used by that particular abstracting service. However, it is entirely possible to pursue a meta-analytic integration of the tests of a specific hypothesis that does not readily fall under any assigned key word in the abstracting service's thesaurus. If the hypothesis under consideration is very specific or very new, the research literature might be too small to have "earned" its own special key word or to have been collectively categorized under any single, more general key word. If the hypothesis under consideration is interdisciplinary, or in any way applied, then untold numbers of tests of the hypothesis may be scattered throughout several different literatures, and fragmentally abstracted by several different abstracting services.

On-line computer searches have the obvious advantages of speed and power; the liability that accompanies these advantages is expense. Depending on the abstracting or indexing services searched, and the level of institutional support, these services can be exorbitantly expensive. In addition, the computer searches carry with them the problems associated with the indexing services (focus on published articles; irrelevance of citations) and the abstracting services (potential inapplicability of the "key words" approach to some content areas).

The invisible college strategy might best be considered a useful and important supplement to the various other strategies for locating studies. No one would claim that the invisible college approach would be sufficient, by itself, to generate a complete and defensible meta-analytic database; in a tight and well-integrated research domain, it might, but in such an area, so would the use of the other strategies. However, despite its potential lack of inclusiveness, the invisible college has the potential to uncover new studies that would be otherwise unavailable.

The limitations of browsing are, of course, an abysmal inefficiency and a completely unsystematic approach. However, like the invisible college, it has the capacity to uncover peripheral studies that might otherwise not be retrieved.

Thus, the various strategies enjoy different strengths and suffer from different weaknesses. The most defensible effort to locate studies for a meta-analytic integration would implement each of the strategies, to take advantage of the strengths of each approach and to compensate for the limitations of each approach. In any event, it is always incumbent on the meta-analytic reviewer to specify the strategies followed in locating studies and to discuss any systematic variants of these types of strategies.

THE "FILE DRAWER PROBLEM"

Rosenthal (1979) identified the "file drawer problem" as the possibility that unknown, unpublished studies might exist, whose results fail to support the pattern established by the published findings. The extreme view of the file drawer problem is stated by Rosenthal (1979) as the possibility that "The journals are filled with the 5% of the studies that show Type I errors [i.e., rejecting a null hypothesis of no difference that is in fact true], while the file drawers back at the lab are filled with the 95% of the studies that show nonsignificant ($p > .05$) results" (p. 638).

In a sense, the file drawer problem can be considered a result of the publication bias in favor of significant results, described earlier in this chapter, in conjunction with an exclusive or inordinate reliance on published research. Although the extreme view of the file drawer problem, as quoted above, is a highly unlikely state of affairs, the point is well taken that a publication bias in conjunction with a retrieval of studies that focuses exclusively on published studies could result in a considerable distortion in the meta-analytic database. There are three approaches to handling the file drawer problem.

"Fail-Safe" Numbers. Rosenthal (1979) demonstrated a means to assess the magnitude of the file drawer problem by calculating the minimum number of unpublished studies showing null results that would be needed to overturn the conclusion reached in the meta-analysis. This number, dubbed the "fail-safe number" by Cooper (1979), estimates the number of unretrieved studies averaging null results, or no effect, that would be needed to bring the overall probability level to the "just significant" $p = .05$. For example, in a meta-analysis of 345 studies of the effects of interpersonal self-fulfilling prophecies, Rosenthal and Rubin (1978) determined that over 65,000 unretrieved studies averaging null results would have to be crammed into the file drawers of researchers somewhere before the overall results could be reasonably ascribed to sampling bias.

Orwin (1983) has proposed an analogous procedure that yields a fail-safe number for effect sizes. This procedure derives the number of unretrieved studies showing null results, or no effect, that would be needed to bring the overall combined magnitude of effect to some negligible level (which would generally be determined by considerations that are specific to the phenomenon being examined, according to this procedure). Although this approach has not attained the currency of the more popular fail-safe number for significance levels, it does present an interesting alternative means of gauging the tolerance of the results of a meta-analytic integration for future null results.

This fail-safe number approach to the file drawer problem provides us with a numerical gauge of the number of unretrieved studies it would take to overturn the results of the meta-analysis. However, there is no standard fail-safe "threshold"; that is, there is no number above which a fail-safe number indicates successful dismissal of the file drawer problem, and below which a fail-safe number indicates that the meta-analytic results may be distorted by the file drawer problem. R. Rosenthal (1984; Rosenthal & Hall, 1981) suggested as a reasonable tolerance level that the fail-safe number exceeds $5k + 10$ (where k = the number of studies included in the meta-analytic database). The $5k$ portion of this benchmark suggests that it is unlikely that there are five times as many studies hidden in the file drawers as included in the reviewer's meta-analytic database; and, the + 10 portion insures that the minimum number of studies that might be hidden in the file drawers is 15, when the number of studies in the data base is one.

The meta-analytic reviewer should recognize that these fail-safe numbers are based on the values obtained from the studies included in the meta-analysis. If these studies are unrepresentative of the general population of studies examining the phenomenon in question, these fail-safe numbers should be viewed with caution (cf. Iyengar & Greenhouse, 1988, Rosenthal & Rubin, 1988, and Strube & Hartmann, 1983, for a critical discussion of the fail-safe number approach).

Funnel Plots. Light and Pillemer (1984) introduced the graphical technique of funnel plots into the meta-analytic arena. Funnel plots take advantage of the well-known *Law of Large Numbers:* In general, the larger the sample size, the more probable it is that the sample mean is a better estimate of the population mean. With reference to study outcomes in a meta-analysis, this suggests that the effects observed in a particular hypothesis test will be a more accurate estimate of the "true" population effects when a larger sample size is used. When study outcomes are plotted as a function of sample size, the operation of the law of large numbers produces a *funnel plot,* so called because the distribution of study outcomes takes the shape of an inverted funnel. As illustrated in Fig. 3.1, study outcomes, presented on an arbitrary metric from −0.5 to +0.5, that are based on relatively small sample sizes exhibit a considerable amount of variability around the mean. However, as sample sizes increase, study outcomes begin to converge on the population mean effect.

Although this is interesting in and of itself, what has this to do with the file drawer problem? The application of funnel plots to the file drawer problem derives from the fact that a truly funnel-shaped funnel plot will not obtain when a publication bias exists. For example, if there is a bias against the publication of nonsignificant results, a funnel plot of the type depicted in Fig. 3.2 might result.

Notice that large study outcomes based on small sample sizes (which

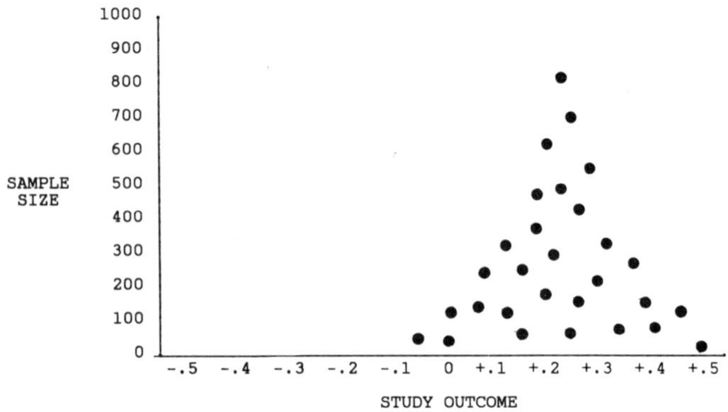

FIG. 3.1. Funnel plot of study outcomes when publication bias does not exist (mean study outcome equals +.25).

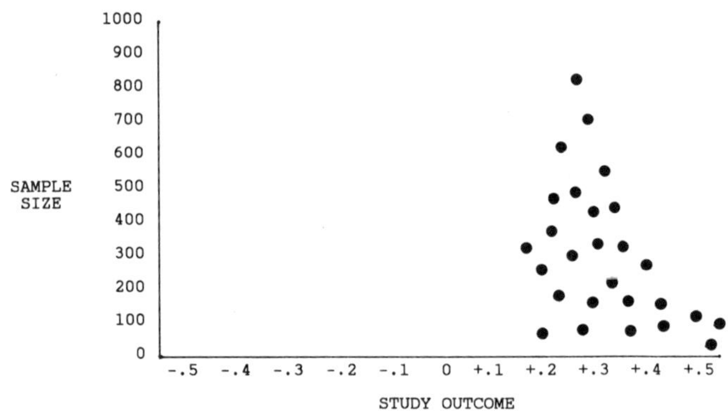

FIG. 3.2. Funnel plot of study outcomes when publication bias does exist (mean study outcome equals +.25).

would still be statistically significant) are represented in this meta-analytic database. In addition, moderate-to-small study outcomes that are based on moderate-to-large sample sizes (which would also be significant) are also represented in this meta-analytic database. However, the portion of the funnel that is missing can be identified as the small study outcomes based on small sample sizes (which would not be statistically significant). Presumably, a publication bias may have been operating here to prevent the publication of the nonsignificant study outcomes, which the law of small numbers dictates would fill up the lower left quadrant of the funnel plot.

Another species of publication bias could evolve when the real typical study outcome is zero. Consider the paradigmatic controversial research domain where some researchers claim to demonstrate that *X* increases *Y*, and other researchers claim to demonstrate that *X* decreases *Y*. If the actual mean effect is zero (i.e., no real effect of *X* on *Y*), and if a publication bias were in operation, we would tend to find that large sample studies, tending to more accurately represent the true typical effect of zero, would not be published, whereas smaller sample studies, tending to produce more extreme (positive or negative) effects, would be published. This pattern, where the "uninteresting" (albeit more accurate) nonsignificant results are not published, would take the form of the "hollow" type of funnel plot presented in Fig. 3.3.

Thus, when the funnel plot takes the shape of a full, inverted funnel, the likelihood of a publication bias and the file drawer problem is reduced. However, when the funnel plot deviates from the shape of a full, inverted funnel, the likelihood of a publication bias and the file drawer problem looms larger and must be addressed.

Just as there was no established, conventional benchmark for the evaluation of fail-safe numbers, there is currently no numerical method for gauging the deviation of a funnel plot from "funnelness". However, two such numerical gauges of funnel plots can be suggested. The distribution that represents the full, inverted funnel should have two numerical characteristics:

1. There should be a strong inverse correlation between sample size and the variability in study outcomes. Looking at the "good" funnel

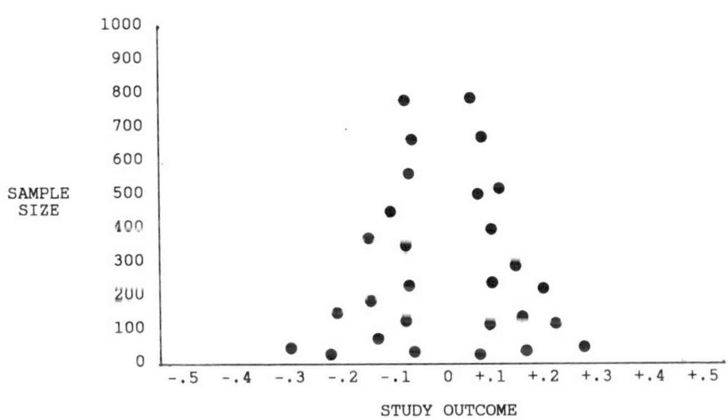

FIG. 3.3 Funnel plot of study outcomes when publication bias does exist (mean study outcome equals 0).

plot of Fig. 3.1, there is very little variability around the typical study outcome for studies with large sample sizes; it is the studies with small sample sizes that differ dramatically from one another.
2. The distribution that represents a "good" funnel is highly peaked, or high in *kurtosis* (cf. Darlington, 1970; Hildebrand, 1971; Moors, 1986, for a discussion of kurtosis as an index of peakedness or bimodality).

Looking at the "good" funnel plot of Fig. 3.1, there is only one modal point in this distribution. One or the other of these two characteristics would be violated by one or the other of the two species of publication bias described earlier. For the first species of publication bias (Fig. 3.2), the absence of studies in the lower left quadrant would dramatically reduce the variability across studies with small sample sizes. For the second species of publication bias (Fig. 3.3), the absence of studies in the center of the inverted funnel would result in a frequency distribution of study outcomes with a gap in the center between two modes. Although these numerical indices can be suggested as preliminary attempts to gauge the "funnelness" of a funnel plot, future efforts will have to delineate the parameters and limitations of these, or other, numerical representations of funnel plots.

Opening the Drawers. Fail-safe numbers and funnel plots can give us an indication of how susceptible our meta-analytic efforts are to the file drawer problem; however, they have about them the sense of "closing the barn door after the horse has bolted." The best solution to the file drawer problem is to open the file drawers. That is, by contacting researchers who are active in the area, by seeking out and obtaining unpublished theses and dissertations, and unpublished convention presentations, the reviewer can maximize the likelihood of obtaining those hypothetical unpublished studies that disconfirm (or complement) the pattern represented in the published findings. If the file drawer problem is a weakness of the meta-analytic approach that results from inadequate study retrieval, then the first line of defense must be a concerted effort to retrieve all potentially relevant studies from all sources. Careful attention to this step in the enterprise will help to insure sufficiently safe fail-safe numbers, and sufficiently funnel-like funnel plots.

4

Retrieving Statistical Tests of the Hypothesis

It's a little like the tale of the roadside merchant who was asked to explain how he could sell rabbit sandwiches so cheap. "Well", he said, "I have to put in some horse meat too. But I mix 'em fifty-fifty: one horse, one rabbit."

—Darrell Huff (1954)

Once the reviewer has developed a clearly specified and well-defined hypothesis test, and the sources of relevant studies have been identified and exhaustively searched for includable studies, the next step is to derive the appropriate statistical tests of the hypothesis under consideration from these studies. Of course, one study may report a *t* test, a second study may report a chi-square, a third study may report a correlation coefficient, and so on. In order to begin to talk about the basic analytic questions of central tendency, variability, and prediction and explanation, these various species of inferential statistics will have to be transduced to a few common metrics. Thus, a critical component of the retrieval of statistical tests of the hypothesis under consideration is expressing the various types of statistics in common metrics.

Occasionally, special circumstances will develop in this process of deriving statistical tests of the hypothesis and transducing these statistics to some common numerical metrics. For example, sometimes, the statistics will be incompletely or improperly reported. Sometimes multiple tests of the same hypothesis will be reported in a single study. And, sometimes, multiple tests of the same hypothesis will come out of a single laboratory. All of these represent special cases of the general process of retrieving hypothesis tests from the studies included in the meta-analysis. After considering each of these issues in turn, we examine a troublesome, but potentially avoidable, oddity about testing hypotheses at the primary level of analysis that may exert unwanted influence at the meta-analytic level.

SIGNIFICANCE LEVEL AND EFFECT SIZE: THE METRICS OF META-ANALYSIS

Up to this point, we have been, purposefully, rather vague about "study outcomes." For example, some of the figures of the previous two chapters depicted an ill-defined "study outcome" that varied as a function of some moderator (in chapter 2), or that was more or less variable as a function of sample size (in chapter 3). However, as we begin to consider how we are going to incorporate and integrate the statistics reported for a particular hypothesis test, we have to become more explicit and more careful about the types of outcomes we can look for in statistical hypothesis tests.

One dimension of study outcome is *significance level.* This refers to the likelihood of having obtained the observed results if, in fact, the null hypothesis of no difference were true. Although we can calculate various types of statistical tests (such as *t*s, *F*s, etc.), the common metrics for significance level are *Z,* or standard normal deviates, and *p values,* or probability values. *Z*s are characterized as having a mean of zero, a standard deviation of one, and as being directly associated with specified *p values.* There is usually a table in statistics textbooks (and it is usually the first table) that lists *Z* scores and their corresponding probability levels. For example, a *Z* of 0.00 corresponds to a one-tailed probability level of $p = .5000$; this represents a lack of any difference between compared conditions, or a lack of any relation between two variables. In other words, 50 times out of 100 (or, just chance) one would find a difference as extreme or more extreme than that found in the obtained results. A *Z* of 1.645 corresponds to a one-tailed probability of $p = .05$; this represents a just significant difference between compared conditions, or a just significant relationship between two variables. In other words, 5 times out of 100, one would find a difference as extreme or more extreme than that found in the obtained results. This ratio is considered, according to convention, to constitute an unreasonably small likelihood (cf. Cowles & Davis, 1982, for a discussion of the origins of the .05 level of statistical significance).

For example, if a researcher obtained a random sampling of five trained salesmen and five untrained salesmen, measured the sales performance of each salesman, and derived the average performance for the two groups, his comparison of the mean performance for trained salesmen and for untrained salesmen might yield an F test of $F(1,8) = 20.259$, $p = .001$. This would indicate that the likelihood of obtaining a difference this large (or larger) is 1 out of 1000, when in fact there was no real difference.

Tables in the backs of statistics books provide critical values or cutoff points for standard and commonly used inferential statistics. These are typically used to evaluate the results of primary-level statistical analyses in an on–off, yes–no fashion: If one's test statistic reaches or exceeds the appropriate cutoff, then one "found something significant"; if one's test statistic fails to attain the appropriate cutoff, then one did not "find something significant." However, one can be much more precise in gauging the probability associated with statistical analyses. Table 4.1 presents the formulae that transduce several common inferential statistics into Zs. In addition, most nonparametric inferential statistics either yield a p value (which can be used to derive Z directly from p–Z tables), or can themselves be transformed into Zs (cf. Siegel, 1956; Siegel & Castellan, 1988). Note that the logarithms presented in these expressions are natural logarithms (i.e., $\log_e$ and not $\log_{10}$).

Figure 4.1 represents the relations between Z, p, and a few common statistical tests, for the situation where $N = 10$. Note that the ts and Fs that correspond to a specific Z and p will vary as a function of sample size.

The other dimension of study outcome is *effect size,* which refers to the magnitude of the effect. Similar to what we see for significance level, the various types of statistical tests can be transduced to a few common metrics for effect size, such as r, Fisher's Z, and d. The product–moment correlation coefficient, r, stands as a very useful and

Table 4.1
Transformation of Inferential Statistics Into Zs

for t: $Z = (df(log(1 + (t^2 / df))))^{1/2} (1 - (1 / (2df)))^{1/2}$
for $F(1,df)$: $Z = (df(log(1 + (F / df))))^{1/2} (1 - (1 / (2df)))^{1/2}$
for χ^2 (1): $Z = \sqrt{\chi^2}$
for r: $t = (r (N - 2)^{1/2}) / (1 - r^2)^{1/2}$ then
$Z = (df(log(1 + (t^2 / df))))^{1/2}(1 - (1 / (2df)))^{1/2}$

Note that the logarithms presented in these expressions are natural logarithms (e.i., $long_e$ and not $\log_{10}$).

p—.500	.200	.050	.001
Z-0.000	0.842	1.645	3.076
$t(8)$-0.000	0.889	1.860	4.501
$F(1,8)$-0.000	0.790	3.459	20.259
χ^2 (1)-0.000	0.706	2.706	5.429

FIG. 4.1. Common metrics for significance levels (N = 10).

accessible index of magnitude of effect. Product-moment correlations range from −1.00 (indicating a perfect inverse relation, where one variable decreases as another increases), to 0.00 (indicating no effect), to +1.00 (indicating a perfect positive relation, where one variable increases as the other increases). One can derive r directly from the various types of inferential statistics, using the formulae presented in Table 4.2.

Although r is easily accessible and readily interpretable, there are problems with using r for computations. For example, the metric of r becomes nonlinear (i.e., the distribution of r becomes skewed) at the extreme ends of the scale. Another way of saying this is that an increase of a certain amount on the scale of r will represent different changes in magnitude of effect at different points along the metric from 0.00 to +/− 1.00. For example, r^2, sometimes called the coefficient of determination, can be thought of as the percent of variability accounted for. An r = .30 indicates that 9% [$(.30)^2$] of the variability in one variable is accounted for by variability in the other variable; an r = .90 indicates that 81% [$(.90)^2$]) of the variability is accounted for. However, consider the consequences of increasing these rs by .05 units. Increasing the r = .30 to r = .35 results in an r^2 = .1225, an increase of 3.25 percent of the variability accounted for. However, the same increase by .05 units from r = .90 to r = .95 results in an r^2 = .9025, an increase of 9.25 percent of the variability accounted for. While we shouldn't rely on this percent of variability accounted for as our functional indicator of effect size, this example helps to illustrate the way in which the metric of r becomes nonlinear at its extreme values (cf. also James, Demaree, & Mulaik, 1986).

Table 4.2
Transformation of Inferential Statistics Into rs

for t: $r = [t^2 / (t^2 + df)]^{1/2}$
for $F(1,df)$: $r = [F / (F + df)]^{1/2}$
for $\chi^2(1)$: $r = (\chi^2 / N)^{1/2}$
for Z: $r = (Z^2 / N)^{1/2}$

Fisher's Z transformation of r, or Z_{FISHER}, represents a solution to this skewness of the distribution of r at extreme values. Similar to the common transformation between Z for significance and p values, the correspondence between Z_{FISHER} and r is commonly tabled in the back of statistics textbooks. The precise transformation is achieved using the following formula:

$$Z_{FISHER} = .5\ \{log[(1 + r) / (1 - r)]\}$$

This Z_{FISHER} can be transformed back into an r using this formula:

$$r = (e^{2Z_{FISHER}} - 1) / (e^{2Z_{FISHER}} + 1)$$

Another way of representing effect size is the standardized difference between the means, represented by d. Conceptually, d represents the difference between the means for two groups, divided by some index of the standard deviation of the measurements involved. For example, a $d = 1.00$ indicates that the mean for one group was one standard deviation above the mean for the other group. There are various ways to derive versions of d, depending on what one defines as the appropriate standard deviation (e.g., dividing the difference between the means by the *SD* for the control group, or by the pooled *SD* derived from both groups), and on whether one adjusts this effect size index for small sample bias (cf. Glass et al., 1981; Hedges & Olkin, 1985). The use of some version of the standardized difference between the means requires decisions as to which of these options are more defensible for a particular research paradigm. Differences of opinion regarding these decisions will yield different indices of effect size for the various species of d (something that does not enter in to the use of r and Z_{FISHER}). There are a few other reasons to prefer the use of r and Z_{FISHER} in performing meta-analytic computations (e.g., the definition of the size of the study varies across different derivations of the t test for d, but not for r; the meta-analyst can directly derive the precise r, and corresponding Z_{FISHER}, from the reported test statistics, whereas d often requires additional information regarding standard deviations and cell sizes, which, if unavailable, must be guestimated; cf. R. Rosenthal, 1984, for a discussion of the comparison between r and d). However, for those who are inclined to think in terms of ds, the standardized difference between the means can be derived directly from r (which was itself derived directly from the test statistics):

$$d = [(4r^2) / (1 - r^2)]^{1/2}$$

This d can be transformed back into an r using this formula

$$r = d / (d^2 + 4)^{1/2}$$

For example, recall the researcher described earlier who tested the difference between two groups of salesmen. The calculated statistic of $F(1,8) = 20.259$, $p = .001$, corresponds to a correlation coefficient of $r = .847$, a Fisher's Z of $Z_{FISHER} = 1.244$, and a standardized difference between the means of $d = 3.183$.

Cohen (1977) identified small, medium, and large effect sizes to correspond to rs of .10, .30, and .50, respectively. Figure 4.2 represents the relations between r, Z_{FISHER}, and d for a few common statistical tests, for the situation where $N = 10$. Note that the ts, Fs, and χ^2s that correspond to a specific r, Z_{FISHER}, and d will vary as a function of sample size.

Thus, the statistical tests of the hypothesis under consideration can be transduced to a few common metrics: Z and p for significance level, and r and Z_{FISHER} for effect size. Recall, from our discussion of the definition of the hypothesis test in chapter 2, that we must be careful to include only test statistics that do constitute tests of a well-defined hypothesis (e.g., F tests with only one degree of freedom in the numerator; χ^2s based on only one degree of freedom). Once we have placed the study outcomes for all includable tests of the hypothesis on these common metrics, we can begin the analytic tasks of determining the typical study outcome, portraying the variability around that typical study outcome, and trying to predict and explain that variability.

SIGNIFICANCE LEVEL VERSUS EFFECT SIZE

Occasionally, scholars will engage in a debate over whether significance levels or effect sizes are more informative or more appropriate (e.g., Becker, 1987; Chow, 1988; Royall, 1986). The demonstration of either

r	.000	.100	.300	.500
Z_{FISHER}	0.000	0.100	0.310	0.549
d	0.000	0.200	0.629	1.155
$T(8)$	0.000	0.284	0.888	1.633
$F(1,8)$	0.000	0.081	0.789	2.667
χ^2 (1)	0.000	0.100	0.900	2.500

FIG. 4.2. Common metrics for effect size ($N = 10$).

point of view has generally taken the form that holding one of these dimensions of study outcome constant, considerable and meaningful changes in the other dimension of study outcome can be demonstrated (thereby implying that the one held constant is less informative).

This debate, and these types of demonstrations, fail to appreciate the elegantly simple mutual dependence of significance level and effect size. As eloquently discussed at length by R. Rosenthal (1984), any test of significance can be expressed as a function of effect size and study size; alternatively, any effect size can be expressed as a function of significance level and study size. These fundamental relationships are depicted in Fig. 4.3. Note that these fundamental relationships provide the templates for the transformations of inferential statistics into *Z*s for significance levels (Table 4.1) and into *r*s for effect size (Table 4.2).

SIGNIFICANCE LEVEL = EFFECT SIZE × STUDY SIZE
EFFECT SIZE = SIGNIFICANCE LEVEL / STUDY SIZE

FIG. 4.3. Fundamental relationships between significance levels, effect sizes, and study sizes.

To illustrate, a significant effect of $p = .049999$ might be of trivially small magnitude if it is based on a large sample of 1000 ($Z_{\text{FISHER}} = 0.052$, $r = .052$, $r^2 = .003$, $d = 0.104$), whereas it might be of large magnitude if based on a small sample of 3 ($Z_{\text{FISHER}} = 1.830$, $r = .950$, $r^2 = .902$, $d = 6.076$). Alternatively, a large effect of $r = .900$ might be nonsignificant if based on a small sample of 3 ($p = .4140$), whereas it might be highly significant if based on a large sample of 1,000 ($p =$ 4.62E-20). The point is that the two study outcomes address two related but different questions about the results of a particular hypothesis test. Significance levels indicate the likelihood that these results are due to chance, and effect sizes indicate the magnitude or strength of these results. Both are informative, in very different ways, regarding different facets of the results of a particular hypothesis test.

SPECIAL PROBLEMS

If every study were to have been conducted completely independently of each other, reporting the complete statistical information for only one test of the hypothesis per study, this chapter on retrieving the statistical tests of the hypothesis would be nearly over. However, most research literatures contain exceptions to this simple, homogenized scenario. These exceptions present some special problems that must be considered.

Data Reconstruction. As we have just seen, the appropriate test statistics can be transformed into a few common metrics of Z and p for significance levels, and r and Z_{FISHER} for effect sizes. However, too often, studies will report "a significant effect ($p < .05$)" or "no significant effect ($p > .05$)." When studies fail to provide a complete report of the necessary statistical information (cf. Oliver & Spokane, 1983), the meta-analytic reviewer may have to reconstruct the statistical tests necessary for inclusion in the meta-analysis. The most accurate method of reconstructing the appropriate statistical tests is to speak to or correspond with the authors of the original research. Researchers often do (and probably should, as a matter of principle) keep files of their old analyses for several years after the publication of the journal article. Personally, I have had the pleasant, if somewhat surprising, experience of having an author successfully retrieve for me supplemental test statistics from a study that was over 20 years old!

However, the meta-analyst should be sensitive to the magnitude and nature of such requests for supplemental statistics. Some researchers become apprehensive about any type of statistical analysis that involves their own published work. Insofar as a meta-analysis requires placing all study outcomes on common metrics (as we have just considered in depth), a meta-analysis has the unique capacity to identify the significance and magnitude of effects of each of the included studies. Understandably, it may be a little unsettling to consider the possibility that one's own research is for some reason producing considerably weaker (or stronger) effects than everyone else's efforts. This may be a serious consideration, particularly in research literatures characterized by controversy or flux. The meta-analyst should simply be just as sensitive to the feelings of the primary-level researchers as the primary-level researcher should be to the feelings of the subjects.

In addition, depending on the primary-level researcher's style of organization, productivity level, and frequency of relocating moves, a request to retrieve the specific t test from a study several years old may actually require a considerable amount of time and effort. Finally, it will often be the case that the original, precise statistics are simply not retrievable (due to fire, flood, divorce, moves, office cleaning, etc.). The point is that obtaining the precise statistical information from the original authors, the most appropriate first step, may require considerable sensitivity, time, and effort, and in fact may not be possible in the end.

The next step would be to try to reconstruct the appropriate test statistics from the available information. For example, consider the following situation: The meta-analyst is conducting an integration of the effects of gender on leadership. A particular primary-level study may have looked at the effect of a particular leadership training program on the leadership behavior of both male and female subjects. However, the primary-level researcher may not have been interested in gender differences, and may have simply mentioned "a significant gender effect, which was not of interest here and will not be considered further in this discussion." The meta-analyst can reconstruct the original, complete ANOVA table (and thereby the unreported main effect for gender, as well as the MS_{ERROR}, to be used for more specific comparisons between genders within levels of training) as long as the original article reports the means and cell sizes for each condition, and one of the F tests (in this instance, either the main effect for training, or the interaction term). This type of reconstruction, though tedious, does allow the retrieval of some statistical tests of the hypothesis that would have been lost forever. Blair Johnson (1989) has developed a fantastically useful software system that facilitates these tedious reconstructions.

Finally, if the original statistics are not retrievable from the authors, and if the research report fails to even convey enough information to reconstruct the appropriate tests, conservative estimates of study outcomes can be made. For example, descriptions of "no significant effect" in the absence of statistics can be conservatively estimated as $p = .50$, $Z = 0.000$. Descriptions of "a significant effect" in the absence of statistics can be conservatively estimated as $p = .05$, $Z = 1.645$. Although these reconstructions are likely to be patently false in any given instance, they are the most defensible (and the most conventional) approximations that can be made in the absence of any statistical test information.

The meta-analytic reviewer should report the number and nature of such reconstructions, and primary-level researchers who have expended their time and effort to provide the reviewer with supplementary statistics should be acknowledged. It is hoped that the growing popularity of meta-analytic procedures will encourage an increase in the precision of statistical reporting in future publications. At the very least, the original test statistic, an accurate p value corresponding to this test statistic, and the relevant sample size or degrees of freedom should be provided (if for no other reason than to allow posterity to most accurately represent the results of a given study in future meta-analyses).

able to create a single "supervariable" (R. Rosenthal, 1984) by averaging the standardized score of each subject on each of the dependent measurements, and then performing the appropriate statistical test on subjects' new supervariable scores. Again, the contributing measurements can be differentially weighted, to reflect differing levels of theoretical relevance. For example, the reviewer could integrate the hypothesis tests derived from self-reports, physiological measures, and verbal disfluencies into a single test of the hypothesis that noise influences anxiety.

Fourth, the reviewer can include all statistics and treat them as if they each represented an independent hypothesis test. In this instance, this fourth option seems least defensible; it violates the assumption of independence, and adds little to the meta-analytic enterprise (except, perhaps, for a little bit of the statistical "noise" that results from including very different operationalizations in the same analysis; cf. chapter 2). For example, the separate tests for self-report, physiological responses, and verbal disfluencies could simply be entered into the analyses as if they were independent.

The second version of this problem is represented by the study that uses several items to measure the same phenomenon several times in the same primary-level study. For example, Sanders and Mullen (1983) used 18 items to measure a particular social projection bias called the false consensus effect (i.e., the tendency to see our own behaviors and opinions as relatively common). These 18 items were administered to the same subjects at the same time, and so are not statistically independent. We can consider the application of the four options just presented to this version of the multiple hypothesis test problem. First, because these items were largely identical in structure and format (the only differences being in the wording of the opinion items), selecting and including the test that is theoretically most relevant seems rather arbitrary and capricious. Second, splitting item types into separate meta-analyses could be defended if these highly similar opinion items could be reliably categorized into separate item types. Third, the results for each separate test can be integrated into a single, global hypothesis test that represents this sampling and measurement procedure. Finally, the results for each item can be included separately, treating each as if it represents an independent hypothesis test.

In point of fact, three of these strategies were implemented in two separate meta analyses of this research domain (Mullen et al., 1985; Mullen & Hu, 1988): separating items types into discrete meta-analyses, combining items into a global hypothesis test for each study, and treating each item separately. It is interesting to note that in both meta-analytic integrations of this phenomenon, the combined level of significance and magnitude of effect were equivalent when items were combined into a single global hypothesis test for each study and when items were treated separately. The treatment of each item as if it were an independent hypothesis test has the added advantage of allowing the meta-analyst to begin to examine theoretically interesting differences between the various items, which get lost when effects are collapsed together (e.g., the size of the actual consensus agreeing with each item; cf. Mullen & Hu, 1988, for a consideration of the trade-offs involved in this approach to the problem).

Clearly, if there are reasons to justify the treatment of separate hypothesis tests derived from the same study as if they were independent, the reviewer should still perform the more defensible (if heavy-handed) analyses that collapse the different items in a given study into a single global hypothesis test. Whatever option is pursued, the reviewer must make explicit the guidelines used in selecting or weighting results derived from multiple dependent variable measures found within a single study.

Multiple Studies From a Given Laboratory. Another source of nonindependence discussed by Strube and Hartmann (1983) arises from multiple studies conducted by a given researcher or a given research team. For example, Strube and Garcia's (1981) meta-analysis of the research examining Fiedler's (1967) contingency theory of leadership behavior found that stronger support for the theory was demonstrated in the studies to come out of the laboratories of Fiedler and his colleagues than in the studies conducted by independent researchers. Similarly, Eagly and Johnson (in press) found greater support for the Petty and Cacioppo (1981) Elaboration Likelihood Model's treatment of involvement in persuasion in studies to come out of the laboratories of Petty and Cacioppo than in studies conducted by independent researchers.

Why do such patterns occur? Similarity of results among such related research efforts may be due in part to common subject populations, common experimental paradigms, more careful development of dependent measurements, or other attributes that are peculiar to a given laboratory. It should be recognized that it is not always the case that a theorist's own studies will provide the strongest (or the only) support for the hypothesis being tested. For example, Mullen (1985) conducted a meta-analysis of Social Impact Theory's (Latane, 1981) predictions regarding the effects of source strength and immediacy in group contexts. The weak and heterogeneous findings in this research domain were not significantly stronger in studies conducted by Latane and his students than in studies conducted by independent researchers.

When multiple studies have been conducted within a given research laboratory, the reviewer might be well advised to separate studies conducted by different research teams. Separate meta-analyses conducted upon these distinct samples of research might reveal the effects of this possible source of nonindependence.

A PROBLEM OF ANALYSIS

The foregoing discussions provide the meta-analytic reviewer with the tools needed to express study outcomes on a few simple, common numerical metrics, and to address the problems of nonindependence. However, there remains one final problem in retrieving study outcomes that is at the same time more subtle and more insidious. Even before the meta-analytic reviewer begins to search for and locate studies, decisions made at the level of the primary research analysis may introduce statistical "noise" into the meta-analytic data base. Throughout our considerations, we have assumed, and will continue to assume, that the primary-level researchers have accurately and honestly reported their inferential statistical tests of the hypothesis under consideration. However, even when this is the case, problems can arise.

Consider the data presented in Table 4.3. Twenty observations are presented with numbers representing the value of *X* (some independent variable) and *Y* (some dependent variable). The hypothesis under consideration is that increases in *Y* are associated with increases in *X*.

Suppose one researcher collects these data and makes the reasonable assumption that an *X* less than 15 can be identified as a low *X*, and an *X* greater than 15 can be identified as a high *X*. Similarly, a *Y* less than 15 can be identified as a low *Y*, and a *Y* greater than 15 can be identified as a high *Y*. If the data in Table 4.3 are recast into a simple 2 × 2 frequency table, we produce Table 4.4.

The chi-square associated with this 2 × 2 contingency table is $\chi^2(1) = 12.8$, $p = .00017$. This corresponds to an effect size of $Z_{FISHER} = 1.099$, $r = .800$, $r^2 = .640$, $d = 2.667$.

Suppose a second researcher collects exactly the same data. However, this second researcher performs the dichotomous split on the *X* variable alone, leaving the *Y* variable as a continuous metric. This second researcher could derive the mean *Y* for the 10 observations with a low *X*

Table 4.3
Data for 20 Observations of *X* and *Y*

Observation	*X*	*Y*
01	10	10
02	10	11
03	11	12
04	11	11
05	11	11
06	12	12
07	12	13
08	13	14
09	13	12
10	14	16
11	16	13
12	17	17
13	17	16
14	18	18
15	18	17
16	19	17
17	19	19
18	19	18
19	20	19
20	20	20

($\overline{Y} = 12.2$), and compare it to the mean *Y* for the 10 observations with a high *X* ($\overline{Y} = 17.4$). The *t* test comparing this two means is $t(18) = 6.6141$, $p = .00000174$. This corresponds to an effect size of $Z_{\text{FISHER}} = 1.227$, $r = .842$, $r^2 = .708$, $d = 3.118$.

Finally, suppose a third researcher collects exactly the same data. However, this third researcher does not perform a dichotomous split on either of the two metrics. This third researcher could derive a correlation between the *X* measurements and the *Y* measurements of $r(18) = .9503$, $p = .000000000122$ (or, $p = 1.22\text{E}-10$). This corresponds to an effect size of $Z_{\text{FISHER}} = 1.835$, $r = .9503$, $r^2 = .903$, $d = 6.105$.

The point is that the very same data set could be used to test the very same hypothesis in distinctly different ways, depending on the assumptions made and the analyses conducted by the primary-level researcher.

Table 4.4
Data from Table 4.3 cast into a 2 × 2 contingency table

	Low X	*High X*
Low *Y*	9	1
High *Y*	1	9

The magnitude of the effect derived from this data might be entered into the meta-analysis as $r = .800$, $r = .842$, or $r = .9503$, depending on how the data is handled at the primary level. It must be recognized that any one of these effect sizes, produced from any one of these analytic strategies, may not be more or less correct than any other. We could generate several reasons why one, both, or neither of the variables should be dichotomized for a particular analysis, within a particular research paradigm.

As the meta-analytic reviewer is trying to retrieve statistical tests of the hypothesis from the research literature, how can this problem be solved? Obviously, in some cases, the problem will never arise in the first place. For example, some research paradigms will exclusively involve the reporting of correlation coefficients; other research paradigms will exclusively involve the comparison between group means, involving t tests or simple F tests. However, in a research literature where different analytic strategies are commonplace at the primary level of analysis, the reviewer has the same two general options open here as everywhere else: focus on the theoretically most illuminating analytic strategy; or, include everything *and* test for differences between analytic strategies. In either case, the reviewer(s) should make explicit which option they have pursued in a particular meta-analysis, and justify this decision.

5

Retrieving Predictors of Study Outcomes

People will not look forward to posterity who never look backward to their ancestors.

—Edmund Burke (1790)

So far, we have considered the careful specification of a well-defined hypothesis, the various strategies for locating studies, and the techniques for retrieving statistical tests of the hypothesis and putting them all on some common metrics. We are nearly ready to begin considering the statistical analyses that address the questions of central tendency, variability, and prediction and explanation. However, in preparation for the analytic questions of prediction and explanation, we must consider the retrieval of predictors and moderators of study outcomes. While the meta-analyst is working through the includable studies, deriving the statistical tests of the hypothesis under consideration, he or she will also be tabulating, estimating, and tracking down indices of variables that might (or, according to some theoretical perspective, should), capture some of the variability in study outcomes. The goal at this stage is to identify, and derive meaningful measures of, variables that can help to predict variation in the significance and magnitude of the hypothesis tests included in the meta-analytic data base.

The phenomenon being addressed by the hypothesis test under consideration will determine what types of variables should be coded as potential predictors of study outcomes. It becomes very important at this stage that the meta-analyst have a very good grasp of the relevant theoretical accounts, and the current debates, regarding the phenomenon being addressed by the hypothesis test. The most valuable and informative predictors may be those predictors that shed some light upon, and cast into sharp relief, the explanatory power of theoretical accounts for the phenomenon under consideration.

For example, one study attribute that can be readily derived from most published reports is the percent of the sample that was male. Simply because it is easy to code for "percent sample male," this is not sufficient reason to expend the effort and time to tabulate this type of study attribute in every meta-analysis. Certainly, it may be appropriate to tabulate "percent sample male" if the meta-analyst has some suspicion that gender differences may exert some previously unspecified influence on the phenomenon under consideration. However, coding each hypothesis test for "percent sample male" becomes truly informative and enlightening if there is a sound theoretical rationale that provides an a priori explanation for gender differences in the significance and magnitude of effect for the hypothesis tests under consideration.

The point is not that the coding of study attributes for exploratory purposes should be discouraged; if anything, it should certainly be encouraged. However, the coding of study attributes for exploratory purposes should not loom so large as to subvert or supplant the retrieval of theoretically informative predictors that can speak to the broader theoretical issues that are directing research in that domain. This is where the real "payoff" lies in the process of retrieving predictors of study outcomes: deriving indices of predictor variables that have the potential to illuminate underlying mechanisms and to address the fundamental theoretical issues in the literature at the time of the review.

We can imagine four general sources of potential predictors of study outcomes. These four sources are presented roughly in order of increasing difficulty and time investment for the meta-analyst (and, perhaps revealingly, in order of decreasing popularity in meta-analyses to date).

DIRECT CODINGS

Direct codings are quantitative attributes that are readily available from the written report. In a sense, these attributes can be "directly coded," or simply tabulated, from the method and procedure sections of the study. Examples include year of publication, percent of sample male, whether the experimenter was present or absent during the procedure, and average age of subjects. For example, Hall (1978) meta-analytically integrated previous research on sex differences in decoding nonverbal cues. More recent studies were found to exhibit larger sex differences; that is, women are generally better than men at decoding nonverbal cues, but this tendency was greater in more recent studies. Hall suggested that improvements in measurement instruments and more powerful primary-level data analyses in the more recent studies may account for this pattern. In fact, this may represent a relatively common, if seldom noticed, phenomenon in the development of research paradigms: The observed magnitude of effect may gradually increase over time as a function of refining our operationalizations and analyses to more effectively manipulate the purported cause and to more sensitively measure the purported effect.

Given the easy access we have to these types of predictors, it is not surprising that these direct coding predictors are the predictors most commonly used in meta-analyses. And, again given their accessability, it is not surprising that these direct codings are often used for exploratory purposes.

JUDGES' RATINGS

A considerably richer, and slightly more complicated, source of predictors is judges' ratings of theoretically relevant variables. Identification of acceptable judges is a critical facet of this approach. One might rely on the judgments of professionals who design and conduct the type of research included in the meta-analysis. Alternatively, one might rely on the judgments of university freshmen who are similar to the subjects studied in the research included in the meta-analysis. Procedurally, the critical facets of the procedure and methods sections of the included studies are presented to the judges, with no information about the authors or results of the studies. Then, the judges assign a number to each study's procedure representing the extent to which a given predictor would be engaged or involved in each study. Several judges typically are used, each judge evaluating each study on several theoretically relevant predictors. The judges' mean rating on a given predictor might then be used to predict and explain differences between study outcomes.

There are a few techniques that help to determine whether we should place any confidence in these judgments. First and foremost, if these judgments are to be trusted, there must be some indication that the judges can agree in assigning numbers to the procedures of each study.

In other words, the judges' ratings must have an acceptable degree of *reliability*. No matter how theoretically relevant and useful a particular variable might seem, if the judges do not agree in the assignments of a numerical index of this variable to the studies' procedures, there is no good reason to use those judgments.

The most direct gauge of the reliability of judges' ratings is the mean correlation between judges' ratings of a given predictor, averaged across the different judges. For example, suppose three judges evaluate the anxiety that would have been produced by the procedure of 10 different studies. Judge A's ratings and Judge B's ratings correlate $r = .75$; Judge B's ratings and Judge C's ratings correlate $r = .80$; and Judge C's ratings and Judge A's ratings correlate $r = .90$. Recall that, in chapter 4, we considered how computations involving correlation coefficients are typically performed on Fisher's Z transformation. The average correlation between judges' ratings is derived by transforming the $[n\ (n - 1) / 2] = 3$ correlations into the corresponding Z_{FISHER}s, deriving the average Z_{FISHER}, and then transforming this mean Z_{FISHER} back into the corresponding mean correlation. Thus, the mean Z_{FISHER} derived for these three interjudge correlations is:

$$\overline{Z}_{FISHER} = (0.973 + 1.099 + 1.472) / 3 = 1.181$$

which, in turn, corresponds to an average correlation coefficient of $\bar{r}(8) = .828$.

The well-known Spearman-Brown result (Walker & Lev, 1953) adjusts this mean interjudge agreement as a function of the number of judges involved, producing what is called an *effective reliability* (R), using the following formula:

$$R = (n\bar{r}) / [1 + (n - 1)\ \bar{r}]$$

where n = number of judges
$\bar{r}$ = mean interjudge correlation

In the present example, the effective reliability based on these three judges amounts to $R = .935$, which is a relatively high degree of interjudge agreement. R. Rosenthal (1984) has developed a table of the effective reliabilities derived from various permutations of mean interjudge correlations and numbers of judges.

Demonstrating reliability in judges' ratings of theoretically relevant predictors is a critical first step. However, even if the judges exhibit a high degree of consensus in their evaluations of the theoretically relevant study characteristics, there is still the distinct possibility that the judges are all (unanimously) incorrect. It becomes critically important to demonstrate that judges' ratings are measuring what they're supposed to measure. In other words, the judges' ratings must have an acceptable degree of *validity*.

Some studies in the meta-analytic data base may have obtained *manipulation checks*. In other words, some studies may have asked pointed questions of the subjects about the situation, the manipulations, or the experimental conditions; responses to these measures can be used to check the effectiveness of the manipulations. For example, of the 10 studies of the effects of socially induced anxiety on performance, 5 studies may have measured subjects' self-reports of anxiety under the experimental, anxiety-producing conditions and under the control, no-anxiety conditions. Each of the five studies would have reported some test statistic comparing the mean self-report of anxiety in experimental conditions with that in control conditions. A compelling indication of the validity of judges' ratings is provided if the judges' ratings correlate with the magnitude of the manipulation checks for the studies that report manipulation checks. The studies that produced stronger differences between experimental and control groups on the manipulation checks should also have received the highest ratings of anxiety induction from the judges; the studies that produced weaker differences between experimental and control groups on the manipulation checks should also have received the lowest ratings of anxiety induction from the judges.

If the judges' ratings agree over all the included studies (i.e., reliability) *and* if the judges' ratings agree with the original subjects' perceptions for those studies that reported manipulation checks (i.e., validity), we derive even more confidence in the accuracy and usefulness of the judges' ratings. Of course, it would be the ideal case where every study included in the meta-analysis reported statistics for a manipulation check as well as for the primary dependent variable of interest. In such an instance, the judges' ratings could be validated against manipulation checks for all possible tests of the hypothesis. However, manipulation check statistics will often be reported for only a subset of studies; and sometimes, of course, they will not be available for any studies. If manipulation check statistics are available for only a subset of the total meta-analytic data base, the reviewer will want to insure that the judges' ratings (which have been validated on this subset of studies) exhibit the same (positive, negative, or absent) relation with study outcomes for both subsets of studies: those that report manipulation checks and those that do not. This separate testing of the predictive power of judges ratings, in conjunction with the validation of judges' ratings against manipulation check effects, and the demonstration of robust interjudge correlations, provides a fairly compelling constellation of evidence that judges' ratings are reliably and meaningfully capturing the theoretically relevant variable of interest.

Note that this elaborate procedure for insuring the reliability and validity of judges' ratings does not mean that these ratings will necessarily predict study outcomes: Reliable and valid judgments of the degree of anxiety induction produced by differing experimental procedures may simply not be empirically related to the magnitude of the effect of the social setting. The point is, we would have more confidence in the failure to demonstrate any pattern of study outcome as a function of judges' ratings if we have been sufficiently careful in testing and refining our judgment procedures.

An early example of the use of judges' ratings is found in Smith and Glass's (1977) meta-analysis of psychotherapy outcome studies. The dependent measures of psychotherapy outcome for each hypothesis test were rated on a 5-point scale of reactivity, or "fakeability," where 1 represented low reactivity (e.g., physiological indices that were relatively insensitive to the goals and needs of the therapist or client) and 5 represented high reactivity (e.g., therapists' ratings of the client they were treating). Psychotherapy outcome was demonstrated to vary as a function of this reactivity of outcome measure. Reliability of these judges' ratings was addressed in a rudimentary fashion by demonstrating that six judges' ratings of a sample of 20 studies (from the total of 375 studies) agreed more than 90% of the time. Percent agreement between judges is interesting, but it does not represent as precise and as meaningful an index of reliability as the mean interjudge correlation and the Spearman-Brown effective reliability. Smith and Glass provided no evidence regarding the validity of these judges' ratings.

Alternatively, Carlson and Miller (1987) conducted a meta-analytic integration of the research examining the effects of mood on helping behavior. Three judges evaluated a total of 85 study procedure and method sections on such theoretically relevant variables as sadness, self-absorption, guilt, responsibility, anger, and frustration. Carlson and Miller reported the following patterns: Judges' ratings of these predictors were generally highly reliable (reliability coefficients ranging between .50 and .93). For those studies reporting the results of manipulation checks, judges' ratings of each predictor correlated with the corresponding manipulation check, and did not correlate with manipulation checks for other, irrelevant predictors. And, the magnitude of helping behavior was predicted by some, but not all, of the theoretically relevant predictors (see Carlson & Miller, 1987, and Miller, 1988, for a discussion of this complex pattern of results). This particular integration stands as one of the more comprehensive and sophisticated applications of the strategy of using judges' ratings of theoretically relevant predictors in meta-analysis.

Quality of Research. One particular dimension that has received a considerable amount of attention as a study attribute susceptible to treatment by judges' ratings is the quality of the studies in the meta-analytic database. For example, the exchange between Smith and Glass (1977; Glass & Smith, 1978) and Eysenck (1978) regarding the controversial 1977 integration of research on psychotherapy highlighted the possible impact of including studies of varying quality. Eysenck (1978) invoked the old axe, "garbage in–garbage out" (p. 517), to characterize the meta-analytic enterprise that aggregates together studies of poor quality with studies of good quality.

If this is a serious concern, then studies can always be disaggregated into subgroups of the meta-analytic database that are defined in terms of differing levels of study quality. The problem is that "study quality" may be much more diffuse than many of the other dimensions that judges might successfully gauge for us. Several investigations have reported what would, at best, be described as moderate interjudge agreement on judgments of the quality of research (e.g., Cicchetti & Eron, 1979; Cooper, 1983; Gottfredson, 1978; Marsh & Ball, 1981; Scarr & Weber, 1978; Stock et al., 1982). Stock et al. (1982) reported that the reliability of judgments of the quality of research included in a meta-analysis could be improved by: using multiple dimensions of study quality; having descriptions of the dimensions printed on the coding sheets used in recording study attributes; and by holding training and practice sessions before and during the interval when the study attributes are being recorded.

Cooper (1984) characterized two general approaches to the coding of quality of research. The *threats to validity* approach requires the reviewer to tabulate the presence or absence (or, possibly, the degree) of each one of a number of specific threats to the validity of the research. Different methodologists have proposed different taxonomies of threats to validity; Cook and Campbell (1979) proposed a fairly comprehensive list of 33 threats to validity that fall within four distinct types: *internal validity* refers to the direct correspondence between experimental treatment and the measured effect; *external validity* refers to the generalizability of the observed results; *construct validity* refers to the possible contamination of operationalizations by other, irrelevant constructs or variables, and *statistical conclusion validity* refers to the power and the appropriateness of the data analyses performed at the primary level of analysis. Meta-analysts who are interested in pursuing the threats to validity approach to quality of research might be best advised to attempt an extensive coding of these types of methodological, procedural, and statistical weaknesses.

The other approach described by Cooper (1984) is the *methods description* approach, which essentially requires the reviewer to tabulate the presence or absence (or, possibly, the degree) of each of a number of specific procedural details about the study. Following from standard methodological sourcebooks like Campbell and Stanley (1963), the reviewer would code for various aspects of design, operationalization, sampling, and statistical analysis. Distinct from the threats to validity approach, which scrutinizes each procedure for things done wrong, the methods description approach scrutinizes each procedure for things done right (or, at least, things done in this way rather than that way).

Glass et al. (1981) examined the results of 11 separate meta-analyses, each of which considered the possible effects of quality of research on study outcomes. Low-quality studies were shown to have the largest effect size; high-quality studies were shown to have moderate effect sizes; and medium-quality studies were shown to have the smallest effect sizes. Although it is not clear what this pattern means, or whether this pattern will obtain in any future meta-analyses, this pattern does indicate that there is no simple linear relation between quality of research and study outcome.

Given the considerably greater time and effort involved in deriving sound and defensible judges' ratings, it should not be surprising that these types of predictors are used less often than the simpler direct codings.

POST-HOC THEORETICAL INDICES

This strategy is less commonly applicable than the direct coding or the expert judgment strategies detailed heretofore. Some research literatures will lend themselves to this approach, and others will not. This approach is conceptually more complicated than the use of direct codings and judges' ratings (although, procedurally, it may be easier than the careful and defensible use of judges' ratings, and almost as easy as direct coding). Sometimes, for some research literatures, there may be something about the procedure, the stimulus materials, or the measurement technique that can be used to derive indices relevant to some basic theoretical question. In other words, there is some numerical facet of the study's procedure that, when subjected to some theory-driven transformation, gives rise to a new variable that can then be used to predict and explain variation in study outcomes.

Note that this approach is similar to the direct coding, in that it does not rely on human judges to estimate the level of some theoretically relevant variable. At the same time, this approach is similar to the use of judges' ratings, in that the reviewer is deriving a number to represent the study's procedure that was not directly presented in the written report. Although this theory-driven transformation is explicit and objective and replicable by anyone who cares to do so, it is likely that this particular facet of the research paradigm was not considered in this particular way in the primary level study.

A couple of examples will help to clarify this uncommon but potentially powerful approach. Mullen, Salas, and Johnson (1988) conducted a meta-analysis of research on the effects of communication network structure, with an eye toward identifying the critical component of network centrality. In the 25 years since social psychology has largely lost interest in the effects of communication network centrality, developments among graph theoretical scholars have generated several distinct ways of conceptualizing centrality in a network: *Degree* (the number of channels of communication open to a given position, signifying the opportunity for communication), *Betweenness* (the probability that others in the network must communicate through a given position, signifying the potential for control of communication), and *Closeness* (the distance from a given position to all other positions in the network, signifying efficiency and independence). The identifications of the network structures used in a particular study (e.g., "a five-person wheel," "a four-person chain") could be subjected to the algorithms that have been developed to gauge the degree, betweenness, and closeness of positions in any communication network. It was found that betweenness was the critical component of network centrality that predicted the magnitude of the effects of performing in communication networks of different structures. Moreover, contrary to the conclusions derived from narrative literary reviews (e.g., Shaw, 1964), the predictive accuracy of this component of network centrality was not moderated by study attributes such as task complexity or knowledge of network structure.

An analogous implementation of post-hoc theoretical indices in meta-analysis was developed in Mullen and Johnson's (in press) integration of research on illusory correlation in stereotyping. This research paradigm presents stimulus materials that vary in the number of exemplars of

Group A and Group B engaging in positive or negative behaviors. The configurations of these stimulus materials were used to derive theoretically defined judgment strategies that had been studied in the broader research literature on covariation judgment in cognitive psychology. Analyses revealed that a paired-distinctiveness strategy (i.e., relying on the proportionate frequency with which the smaller group engaged in the rarer behavior) was a very good predictor of the degree of stereotyping.

The point is that these types of predictor variables were implicit and hidden within the procedure sections of all of the previously published research. On the one hand, these post-hoc theoretical indices were not directly codable from the written study reports; on the other hand, these post-hoc theoretical indices did not require any subjective judges' ratings (of problematic reliability and validity). Although the use of this type of post-hoc theoretical index requires a considerable amount of theoretical legwork before the implementation of the meta-analysis, this approach has much to recommend it as a means of trying to predict and understand variability between study outcomes.

PREDICTORS DERIVED FROM ARCHIVAL AND HISTORICAL SOURCES

This approach, even more so than the post-hoc theoretical indices strategy, requires the best detective skills at the disposal of the meta-analytic reviewer. And, again like the post-hoc theoretical indices, some research literatures will lend themselves to this approach, and other research literatures will not. Sometimes, for some research literatures, there may be important outside sources of information about each of the included studies that serves to illuminate differences between studies. Without relying on direct codings of study attributes, or human judgments about certain characteristics of the studies, or theory-driven transformations of attributes of the studies' procedural details, one might be able to locate information about the sample, setting, or historical context of each study that helps to explain why some studies obtained larger and more significant effects than other studies.

For example, Mullen and Felleman (in press) conducted a meta-analytic integration of the research on the effects of "tripling" in college dormitories (i.e., the practice of assigning three people to live in a college dormitory room designed to accomodate two people). Comparisons of social, effective, behavioral, and residential responses of people living two to a room were invaribly more positive than those of people living three to a room of the same size. One phenomenon in the context of the broader literature on crowding in general is the tendency for longer corridors to exert the oppressive, crowding effects often associated with increased interpersonal density (cf. Baum et al., 1978). In an application of the use of historical and archival sources to obtain theoretically relevant predictors, phone calls to the original authors of the primary-level investigations (and sometimes to their former research assistants) allowed us to reconstruct the list of college dormitories on each campus where the original studies were conducted. Then, numerous phone calls and letters later, we were able to obtain the blueprints and floor plans for each of the relevant college dormitories from various Deans of Student Affairs scattered across the country. An intriguing pattern emerged: While students living three to a room generally exhibited more adverse reactions than students living two to a room, this general pattern was exacerbated by corridor length. In other words, the bad effects of tripling were made even worse by longer corridors. This result demonstrated, for the first time, the potential interaction between two distinct manipulations of crowding.

A similar illustration can be found in Mullen, Copper, and Driskell's (in press) meta-analytic integration of the effects of model behavior on jaywalking in pedestrians. Generally, obedient (nonjaywalking) models decreased pedestrian jaywalking, and disobedient (jaywalking) models increased pedestrian jaywalking. An archival source (Flubert, 1976) was located that tabulated the percent of pedestrian traffic flow to occur during each hour of the day. This, in conjuction with each study's report of the time of day during which data was collected, was used to predict the effects of crowded sidewalks on the impact of obedient and disobedient models. These analyses revealed that crowded sidewalks exaggerated the tendency for a disobedient model to increase jaywalking, and that crowded sidewalks undermined the tendency for obedient models to decrease jaywalking. These results bear upon several theoretical accounts for the effects of crowding on behavior.

Thus, these two meta-analyses illustrate the successful employment of historical and archival sources to try to account for theoretically interesting variability in study outcomes. Again, this approach does not involve the direct codings of information from the written reports of the studies; it does not involve any judges' ratings; and, it does not even involve the post-hoc derivation of theoretically relevant variables from procedural information conveyed in the written report. There is no question that this strategy for obtaining theoretically illuminating predictors involves a tremendous amount of effort and tenacity on the part of the meta-analytic reviewer/private investigator. Nonetheless, when a particular phenomenon or research paradigm leads itself to this approach, historical and archival sources present the opportunity to examine theoretical issues that might be prohibitively impractical, or that may have never even been imagined, at the primary level of analysis.

These four strategies for retrieving predictors of study outcomes have complementary strengths and weaknesses. The direct coding approach is certainly the simplest and most straightforward. At the same time, it is probably the most limited in depth and breadth, being restricted to the variables that primary-level researchers saw fit to report in the first place. The use of judges' ratings is slightly more tedious, procedurally, and may be demonstrated to be invalid and/or unreliable in any particular application. However, if demonstrated to have acceptable reliability and validity in a given context, judges' ratings allow the reviewer to ask the kinds of questions that they find interesting (rather than being largely restricted to dealing with the kinds of variables that the primary-level researchers found interesting). The use of post-hoc theoretical indices and historical or archival sources are unquestionably tedious, demanding, and difficult. However, they have the benefits of being particularly compelling, and illuminating of issues and questions that might not be accessible through either of the other two strategies.

6

Combinations of Significance Levels and Effect Sizes

Add little to little and there will be a big pile.

—Ovid

At this step in the meta-analysis, the reviewer has already covered a tremendous amount of ground. A careful definition of the hypothesis was developed, the involved and tedious task of retrieving includable studies has been accomplished, the relevant statistical tests of the well-defined hypothesis have been retrieved from each study, these statistical tests have been transformed to the common metrics of significance levels and effect sizes, and the theoretically interesting predictors have been retrieved for each study.

The reviewer is now ready to begin asking the three general analytic questions of *central tendency, variability,* and *prediction.* Our emphasis in this chapter is on the general analytic question of central tendency: What is the typical study outcome? In particular, what is the combined level of significance, and what is the mean magnitude of effect?

As we consider each of the general analytic questions, we employ a single set of studies to illustrate the calculation and interpretation of these meta-analytic statistics. Suppose that an exhaustive search has uncovered a total of 10 studies that represent specific tests of the well-defined hypothesis that Y increases as a function of X (the reader can substitute for X and Y any variables of interest: treatment and therapeutic improvement; group size and performance; intervention and outcome; etc.). The number of hypothesis tests in the meta-analytic data base is represented as k. Thus, for this particular meta-analytic data base, $k = 10$. For the sake of this illustration, assume that operationalizations of the independent variable X and of the dependent variable Y are defensibly equivalent across all 10 studies.

Let us assume that the reviewer has already retrieved this meta-analytic data base of 10 studies. In addition, let us assume that the reviewer has retrieved the statistical tests of the hypothesis and transformed these hypothesis tests to the common metrics of significance level and effect size. Further, let us assume that each hypothesis test has been coded for three predictors: year of publication; magnitude of the manipulation of X (i.e., the duration of the treatment; the size of the group; the strength of the intervention); and, presence of the observer during data collection (i.e., present vs. absent). Although we do not consider these predictors until subsequent chapters, they are presented here for the sake of completeness. The hypothesis tests, corresponding statistical information, and predictors are presented in Table 6.1.

Note that the statistics derived for each hypothesis test have already been transformed into the common metrics of Z and one-tailed p (for significance level), and Z_{FISHER}, r, r^2, and d (for effect size). These transformations are obtained using the algorithms presented in chapter 4. This type of table, presenting statistical tests, common metrics for significance levels and effect sizes, and predictors, for each hypothesis test, should be presented in any meta-analysis (Bullock & Svyantek, 1985). Notably, this provides useful information to the interested reader. However, it also provides necessary information for the reviewer or critic who may wish to pursue alternative meta-analytic strategies (cf. Fiske, 1983).

As we begin to consider the first analytic question of characterizing the typical study outcome, or the *central tendency* of the results obtained in these 10 hypothesis tests, we perform our computations on the two sets of Zs: the standard normal deviate Z for significance level, and the Fisher's Z transformation of r for effect size.

COMBINATION OF SIGNIFICANCE LEVELS

This meta-analytic procedure allows the meta-analyst to estimate the probability that the p values of the k included studies might have been obtained if the null hypothesis were true. From the statistical test of the hypothesis reported for each study, a Z is obtained for significance level. Each Z is then multiplied by a weight, w. If $w = 1$ is used, this produces

Table 6.1
Hypothesis Tests Included in Meta-Analytic Data Base

Study	Statistic (df)	N	DOE	Significance Effect Size						Year	Mag. of X	OP
				Z	p	Z_{FISHER}	r	r^2	d			
1	$\chi^2 = 23(1)$	110	+	4.80	8.35E−7	0.49	.457	.209	1.03	1981	7	0
2	$t = 6.5(98)$	100	+	5.91	2.08E−9	0.62	.549	.301	1.31	1984	5	0
3	$F = 15.0(1,88)$	90	+	3.71	.00010	0.40	.382	.146	0.83	1985	8	1
4	$r = .335(78)$	80	+	3.04	.00119	0.35	.335	.112	0.71	1982	5	0
5	$p = .000001(-)$	80	+	4.75	.000001	0.59	.531	.282	1.25	1983	8	0
6	$Z = 3.891(-)$	70	+	3.891	.00005	0.50	.465	.216	1.05	1988	3	1
7	$p = .01(-)$	60	+	2.33	.01	0.31	.300	.090	0.63	1990	2	1
8	$F = 10.25(1,63)$	65	+	3.07	.00107	0.39	.374	.140	0.81	1986	4	0
9	$r = .535(68)$	70	+	4.77	9.45E−7	0.60	.535	.286	1.27	1987	10	1
10	$t = 6.0(63)$	65	+	5.31	5.79E−8	0.70	.603	.364	1.51	1989	9	1

Note: DOE = Direction Of Effect (where + = expected direction,
− = unexpected direction)
Mag. of X = Magnitude of manipulation of independent variable X
OP = Observer Presence (where 1 = Present,
0 = Absent)

an unweighted Stouffer's (1949) method of combining probabilities. With an unweighted combination of significance levels, each hypothesis test contributes an equal amount to the combination of significance levels. Alternatively, if some w other than 1 (one) is used, this produces a weighted Stouffer's (1949) method of combining probabilities. Unequal weighting could allow some studies to contribute more than others to the combination of significance levels. For example, studies based on larger samples (if $w = N$ for each hypothesis test) or studies based on a higher procedural and methodological quality (if w = mean reliable judges' ratings of quality of research) could contribute more to the combination of significance levels than smaller sample or poorer quality studies.

A particular research domain may dictate that each study be weighted equally, or that each study be weighted as a function of sample size. Sample size is often a useful and appropriate weight, which may help to better capture the general patterns of results in a given research domain. For example, suppose one wishes to represent the combined probability of the highly significant results of a study based on 10,000 subjects, and the nonsignificant results of a study based on 10 subjects. Giving equal weight to these two studies may dramatically misrepresent the typical probability of the phenomenon under consideration. If it is not self-evident which weighting procedure is more appropriate for a particular research domain, the reviewer should conduct, and report the results of, both weighted and unweighted analyses.

The formula for the meta-analytic procedure of combining significance levels is:

$$Z = \frac{\Sigma w_j Z_j}{\sqrt{\Sigma w_j^2}}$$

where: w_j = weight assigned to results of hypothesis test j
Z_j = Z associated with significance level of hypothesis test j

The p value corresponding to this new Z is the likelihood that the combined results of the k studies, or a result more extreme, would have been obtained if the null hypothesis were true.

Another aspect of the combination of significance levels involves the calculation of the number of unretrieved studies averaging null results needed to bring the overall combined probability to some specified level. This "fail-safe number" for significance levels was discussed previously in chapter 3 as a response to the file drawer problem. The formula (cf. Rosenthal, 1979) for the fail-safe number for the p = .05 level of significance is:

$$Nfs\ (p = .05) = \left(\frac{\Sigma Z_j}{1.645}\right)^2 - k$$

where: Z_j = Z associated with significance level of hypothesis test j
k = number of hypothesis tests included in meta-analysis

The denominator of 1.645 represents the Z corresponding to the "p = .05" level of significance typically used to determine whether or not a result is considered of negligible significance. This fail-safe number estimates the number of new, filed, or unretrieved studies averaging null results that would be required to bring the overall combined probability level to the "just significant" convention of $p = .05$.

For example, consider the calculation of the combination of significance levels for the $k = 10$ hypothesis tests represented in Table 6.1. The Z for combination of significance is derived as:

$$Z = \frac{110(4.80) + 100(5.91) + 90(3.71) + 80(3.04) + 80(4.75) + 70(3.891) + 60(2.33) + 65(3.07) + 70(4.77) + 65(5.31)}{\sqrt{(110)^2 + (100)^2 + (90)^2 + (80)^2 + (80)^2 + (70)^2 + (60)^2 + (65)^2 + (70)^2 + (65)^2}}$$

$$= 13.221$$

The probability associated with this new Z for combination of significance levels is approximately p = 9.01E-30. This represents scientific notation for p = .00000000000000000000000000000901. Thus, the results of the hypothesis tests reported in Table 6.1 were highly unlikely to have occurred if the null hypothesis were true.

Consider the calculation of the fail-safe number for the $k = 10$ hypothesis tests represented in Table 6.1. The fail-safe number for significance levels for the $p = .05$ level of significance is derived as:

$$Nfs\ (p = .05) = \left(\frac{4.80 + 5.91 + 3.71 + 3.04 + 4.75 + 3.891 + 2.33 + 3.07 + 4.77 + 5.31}{1.645}\right)^2 - 10$$

$$= 628.98$$

Thus, approximately 629 additional studies averaging null results would be necessary to bring the overall combined probability level to $p = .05$. Recall the "5 k + 10" benchmark suggested by R. Rosenthal (1984; Rosenthal & Hall, 1981); with $k = 10$, a tolerable fail-safe number for this database would be set at 60. Since the obtained fail-safe number exceeds this benchmark (by an order of magnitude), we could safely conclude that these results appear to be acceptably tolerant for future null results.

COMBINATIONS OF EFFECT SIZES

This meta-analytic procedure allows the meta-analyst to estimate the mean effect size for the *k* included studies. From the statistical test of the hypothesis reported for each study, a product moment r is obtained for effect size. Then, Z_{FISHER}s can be derived from these *r*s. Analogous to the procedure followed for the combination of significance levels, these Z_{FISHER}s are then multipled by a weight (e.g., "1", *N*, etc.), and the sum of these weighted Z_{FISHER}s is then adjusted by a sum of the weights. Just as with the combination of significance levels, a particular research domain may dictate that each study be weighted equally, or that each study be weighted by sample size. When in doubt, the reviewer should conduct, and report the results of, both weighted and unweighted analyses.

The formula for the meta-analytic procedure of combination of effect sizes is:

$$Z_{FISHER} = \frac{\Sigma w_j Z_{FISHER}}{\Sigma w_j}$$

where: w_j = weight assigned to the results of hypothesis test *j*
Z_{FISHER} = Z_{FISHER} associated with effect size for hypothesis test *j*

This estimates the mean magnitude of effect for the *k* hypothesis tests included in the meta-analysis. From this Z_{FISHER}, the reviewer can work back to derive the corresponding *r*, r^2, and *d*, using the algorithms presented in chapter 4.

Another aspect of the combination of effect sizes involves the calculation of the number of unretrieved studies averaging null results needed to bring the overall combined effect size to some specified level. This "fail-safe" number for effect size was discussed in chapter 3 as a response to the file drawer problem. The formula (cf. Orwin, 1983) for the fail-safe number for effect sizes is:

$$Nfs(Z_{FISHER} = 0.100) = \frac{k(Z_{FISHER} - Z_{FISHER_c})}{Z_{FISHER_c}}$$

where: k = number of hypothesis tests included in meta-analysis
$\overline{Z}_{\text{FISHER}}$ = mean Z_{FISHER} for k hypothesis tests
Z_{FISHER_c} = Z_{FISHER} associated with specified criterion

The denominator of $Z_{\text{FISHER c}}$ represents the Z_{FISHER} corresponding to some magnitude of effect that the reviewer wishes to use to determine whether the result is of negligible effect size. This fail-safe number estimates the number of new, filed, or unretrieved studies averaging null results that would be required to bring the overall combined effect size to the level specified by the $Z_{\text{FISHER c}}$, the specified criterion effect size. Note that, aside from Cohen's (1977) seldom-used benchmarks for small, medium, and large magnitudes of effect, there is no convention for effect size that ever achieved the currency and popularity of the "$p = .05$" convention for significance levels. Therefore, unlike the established conventional criterion built into the fail-safe number for significance levels, there is no established, conventional criterion built into the computation of a fail-safe number for effect sizes.

For example, consider the calculation of the combination of effect sizes for the $k = 10$ hypothesis tests represented in Table 6.1. The mean Z_{FISHER} for combination of effect sizes is derived as:

$$Z_{\text{FISHER}} = \frac{110(0.49) + 100(0.62) + 90(0.40) + 80(0.35) + 80(0.59) + 70(0.50) + 60(0.31) + 65(0.39) + 70(0.60) + 65(0.70)}{110 + 100 + 90 + 80 + 80 + 70 + 60 + 65 + 70 + 65}$$

$$= 0.499$$

This Z_{FISHER} of 0.499 corresponds to $r = .461$, $r^2 = .213$, and $d = 1.039$. Thus, in line with Cohen's (1977) estimates of small, medium, and large effect sizes, the mean effect size for the $k = 10$ studies represented in Table 6.1 is of medium to large magnitude.

Consider the calculation of the fail-safe number for effect sizes for the $k = 10$ hypothesis tests represented in Table 6.1. For the sake of illustration, we can use as the criterion effect size Cohen's small effect of $Z_{\text{FISHER}} = 0.100$. The fail-safe number for effect sizes for the $Z_{\text{FISHER}} = 0.100$ magnitude of effect is derived as:

$$Nfs(Z_{\text{FISHER}} = 0.100) = \frac{10\,(0.499 - 0.100)}{0.100}$$

$$= 39.9$$

Thus, approximately 40 additional studies averaging null results would be necessary to bring the overall combined effect size to $Z = 0.100$. This fail-safe number for effect sizes is more difficult to interpret than the more commonly used fail-safe number for significance levels. The "$5\,k + 10$" benchmark suggested by R. Rosenthal (1984; Rosenthal & Hall, 1981) for the fail-safe number for significance levels may or may not apply to this fail-safe number for effect sizes; the issue has never been addressed. Moreover, again, there is no established convention defining what the minimal criterion effect size should be in deriving such a fail-safe number for effect sizes. Therefore, this approach should be considered merely for illustrative purposes, until the parameters of Orwin's fail-safe number for effect sizes have been more clearly delineated.

FUNNEL PLOTS

Chapter 3 presented Light and Pillemer's (1984) use of funnel plots to determine whether publication bias may be influencing or distorting the results of combinations of study outcomes. Recall that funnel plots take advantage of the law of large numbers to construct a plot of effect sizes by sample sizes that, under ideal conditions, should look like an inverted funnel. Again, the rationale is that there is likely to be a considerable amount of variability in the sample estimates of effect size based on small sample sizes; however, with larger and larger samples, the effect sizes obtained in a given study should more closely approximate the "true" population effect size. One reflection of a publication bias would be a funnel plot that has a large chunk missing in the lower left quadrant of the funnel plot. The hypothesis tests that would fall into the lower left quadrant would be hypothesis tests that obtained small effects based on small samples; these hypothesis tests would tend to be nonsignificant. If the funnel plot reveals that this "small sample—small effect—nonsignificant result" hypothesis test is conspicuous in its absence, this would suggest that there were some factors working against the publication of nonsignificant results in that research domain.

For example, Fig. 6.1 presents the funnel plot derived from the *N*s and the Z_{FISHER}s presented in Table 6.1. This figure indicates a fairly respectable funnel shape, considering the relatively small number of hypothesis tests included in this meta-analytic data base.

However, note that this particular funnel plot, based on these particular 10 hypothesis tests, does not unequivocally rule out the possibility of

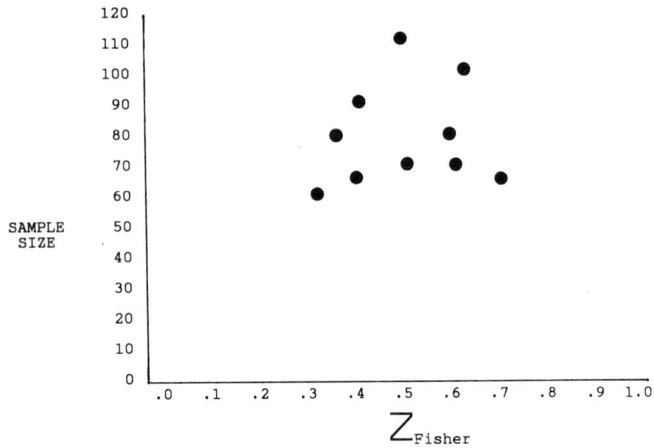

FIG. 6.1. Funnel plot derived from data in Table 6.1.

a publication bias in this particular research domain. This is because every one of the 10 hypothesis tests in Table 6.1 is statistically significant. Although the pattern depicted in Fig. 6.1 has the general shape of a funnel, this particular meta-analytic data base does not include hypothesis tests that would fall into the true lower left quadrant of the funnel plot. What if an eleventh study, based on a sample of 30, obtained an effect size of $r = .212$ (the same magnitude of effect obtained in hypothesis test 7 of Table 6.1)? Such a hypothesis test would yield a significance level of $p = .1304$. This eleventh hypothesis test, and similar hypothesis tests based on smaller sample sizes, would provide more of an acid test for the funnel plot gauge of publication bias. Still, given the small number of hypothesis tests included in this meta-analytic database, the symmetry and the funnelish shape of this plot are consistent with the law of large numbers, and should lend some modicum of confidence to our foregoing combinations of significance levels and effect sizes.

7

Diffuse Comparisons of Significance Levels and Effect Sizes

Perhaps an analyst looking at a group of studies is rather like an artist looking at a complex painting on a canvas. She could look at each of the swirls and brush strokes singly, or she could stand back and try to put each mark in the context provided by its neighbors. We believe that when a social scientist examines variation among studies, it is done in this spirit of building a context.

—Pillemer and Light (1980)

When the primary-level researcher begins to examine the data obtained in a particular study, the first question to be asked is, "What is the typical response," or, "What is the average?". As we saw in the last chapter, this is also the first question to be asked in a meta-analytic integration of a research domain. Thus, in the early stages of a meta-analysis, the reviewer asks, "What is the typical study outcome," or "What is the average effect?".

However, a richer understanding of the phenomenon at hand requires the development of the second general analytic question of variability around the typical outcome. Consider the old axiom from introductory-level statistics classes about the potential inaccuracy of the mean or average score. Suppose a physical education instructor is interested in ordering sweatpants for the 50 males in his weekly gym class. He may determine that the average height of the class is 5′ 10″, and so order 50 pairs of size Medium sweatpants. Of course, if 25 of the students are 6′ 6″ basketball players, and the other 25 students are 5′ 2″ jockeys and gymnasts, that "typical" height of 5′ 10″ in fact characterizes no single individual in the class (and none of the ordered sweatpants would fit any of the students). The same problem exists in dealing with the typical or combined study outcomes at the meta-analytic level. Our emphasis in this chapter is on the general analytic question of variability: How much variability is there around the typical study outcome? In particular, to what extent does the combined level of significance, and the mean magnitude effect, accurately characterize the results of this research domain as a whole?

Note that this type of comparison is referred to as the *diffuse* comparison of significance levels and effect sizes. This is to be distinguished from the *focused* comparison, which is delineated in the next two chapters. At this juncture we can highlight a few major conceptual differences between diffuse comparisons and focused comparisons. As the term implies, the diffuse comparison is rather diffuse, nondirectional, and unspecified, regarding the type of variability in study outcomes. Diffuse comparisons do not reveal anything directly about why study outcomes may differ from one another; diffuse comparisons simply indicate the extent to which there is a considerable amount of spread around the combined or average study outcome. On the other hand, focused comparisons are focused in the sense of focusing on some specific, directional, well-defined account for how study outcomes could differ.

Another way to conceptualize this is that the diffuse comparisons are directed toward the second general analytic question of *variability,* whereas the focused comparisons are directed toward the third general analytic question of *prediction* and *explanation.* As we consider diffuse comparison techniques in this chapter, and focused comparison techniques in the next two chapters, bear in mind the functional distinction between these two types of comparisons: A significant diffuse comparison of study outcomes means that there is a considerable amount of variability in the research domain around the typical or average effects obtained in the combination techniques just reviewed. A significant focused comparison of study outcomes means that the predictor used to formulate that specific focused comparison may provide a meaningful and powerful account for the variability in the research domain.

Once again, we employ a given set of 10 hypothesis tests to illustrate the calculation and interpretation of these meta-analytic statistics. The $k = 10$ hypothesis tests used before, along with the corresponding statistical information and predictors, are presented once again in Table 7.1.

Note, once again, that the statistics derived for each hypothesis test have already been transformed into the common metrics of Z and

Table 7.1
Hypothesis Tests Included in Meta-Analytic Data Base

Study	Statistic (*df*)	*N*	DOE	*Significance*		*Effect Size*				Year	Mag. of *X*	*OP*
				Z	*p*	Z_{FISHER}	r	r^2	d			
1	$\chi^2 = 23(-)$	110	+	4.80	8.35E − 7	0.49	.457	.209	1.03	1981	7	0
2	$t = 6.5(98)$	100	+	5.91	2.08E − 9	0.62	.549	.301	1.31	1984	5	0
3	$F = 15.0(1,88)$	90	+	3.71	.00010	0.40	.382	.146	0.83	1985	8	1
4	$r = .335(78)$	80	+	3.04	.00119	0.35	.335	.112	0.71	1982	5	0
5	$p = .000001(-)$	80	+	4.75	.000001	0.59	.531	.282	1.25	1983	8	0
6	$Z = 3.891(-)$	70	+	3.891	.00005	0.50	.465	.216	1.05	1988	3	1
7	$p = .01(-)$	60	+	2.33	.01	0.31	.300	.090	0.63	1990	2	1
8	$F = 10.25(1,63)$	65	+	3.07	.00107	0.39	.374	.140	0.81	1986	4	0
9	$r = .535(68)$	70	+	4.77	9.45E − 7	0.60	.535	.286	1.27	1987	10	1
10	$t = 6.0(53)$	65	+	5.31	5.79E − 8	0.70	.603	.364	1.51	1989	9	1

Note: DOE = Direction Of Effect (where + = expected direction,
− = unexpected direction)
Mag. of *X* = Magnitude of manipulation of independent variable *X*
OP = Observer Presence (where 1 = Present,
0 = Absent)

one-tailed p (for significance level), and Z_{fisher}, r, r^2, and d (for effect size). These transformations are obtained using the algorithms presented in chapter 4. Just as in the previous set of analyses looking at combinations of study outcomes, as we begin to consider the second analytic question of variability around the typical study outcome, we perform our computations on the two sets of Zs: the standard normal deviate Z for significance level, and the Fisher's Z transformation of r for effect size.

DIFFUSE COMPARISON OF SIGNIFICANCE LEVELS

This meta-analytic procedure allows the meta-analyst to estimate the probability that the p values of the k included hypothesis tests differ significantly. In other words, a significant diffuse comparison of significance levels indicates that the significance levels of the included hypothesis tests were significantly heterogeneous and may be thought of as having been sampled from different populations of study outcomes. From the statistical test of the hypothesis reported for each study, a Z is obtained for significance level. Then, the mean Z for significance level is determined for this research domain (note that, strictly speaking, the mean Z for significance level was not involved in the derivation of the combination of significance levels presented in chapter 6). After the mean Z for significance levels has been obtained, the formula for the meta-analytic procedure of diffuse comparison of significance levels is:

$$\chi^2_{(k-1)} = \Sigma(Z_j - \overline{Z})^2$$

where: Z_j = Z associated with significance level of hypothesis test j
$\overline{Z}$ = mean Z
k = number of hypothesis tests included in the meta-analysis

The p value corresponding to this χ^2 on $(k-1)$ degrees of freedom is the likelihood that the k study outcomes are hetereogeneous, or significantly different from one another.

For example, consider the calculation of the diffuse comparison of significance levels for the $k = 10$ hypothesis tests represented in Table 7.1. First, the mean Z for significance levels is derived as:

$$\overline{Z} = \frac{(4.80 + 5.91 + 3.71 + 3.04 + 4.75 + 3.891 + 2.33 + 3.07 + 4.77 + 5.31)}{10}$$

$$\overline{Z} = 4.158$$

The χ^2 for diffuse comparison of significance levels is derived as:

$$\chi^2 = (4.80 - 4.158)^2 + (5.01 - 4.158)^2 + (3.71 - 4.158)^2 + (3.04 - 4.158)^2 + (4.75 - 4.158)^2 + (3.891 - 4.158)^2 + (2.33 - 4.158)^2 + (3.07 - 4.158)^2 + (4.77 - 4.158)^2 + (5.31 - 4.158)^2$$

$$= 11.6$$

The probability associated with this χ^2 based on $(k - 1)$ 9 degrees of freedom is $p = .2358$. Thus, the significance levels of the 10 hypothesis tests reported in Table 7.1 are not significantly different from one another.

DIFFUSE COMPARISONS OF EFFECT SIZES

This meta-analytic procedure allows the meta-analyst to estimate the probability that the effect sizes of the k included hypothesis tests differ significantly. In other words, a significant diffuse comparison of effect sizes indicates that the strength of effects of the included hypothesis tests were significantly heterogeneous and may be thought of as having been sampled from different populations of study outcomes. From the statistical test of the hypothesis reported for each study, a r for effect size is obtained, and Z_{FISHER}s are, in turn, derived from these rs. Then, the weighted mean Z_{FISHER} for effect size is determined for this research domain. Note that this mean Z was already obtained in the combination of effect sizes presented in chapter 6 (*if* the Z_{FISHER}s were weighted by sample size). The formula for the meta-analytic procedure of diffuse comparison of effect sizes is:

$$\chi^2_{(k-1)} = \Sigma(N_j - 3)\,(Z_{FISHER} - \overline{Z}_{FISHER})^2$$

where: N_j = N associated with hypothesis test j
Z_{FISHER_j} = Z_{FISHER} associated with hypothesis test j
$\overline{Z}_{FISHER}$ = mean Z_{FISHER}
k = number of hypothesis tests included in the meta-analysis

The p value corresponding to this χ^2 on $(k - 1)$ degrees of freedom is the likelihood that the k study outcomes are heterogeneous, or significantly different from one another.

For example, consider the calculation of the diffuse comparison of effect sizes for the $k = 10$ hypothesis tests represented in Table 7.1. The χ^2 for diffuse comparison of effect sizes is derived as:

$$\begin{aligned}\chi^2 = {} & (110 - 3)(0.49 - 0.50)^2 + (100 - 3)(0.62 - 0.50)^2 + \\ & (90 - 3)(0.40 - 0.50)^2 + (80 - 3)(0.35 - 0.50)^2 + \\ & (80 - 3)(0.59 - 0.50)^2 + (70 - 3)(0.50 - 0.50)^2 + \\ & (60 - 3)(0.31 - 0.50)^2 + (65 - 3)(0.39 - 0.50)^2 + \\ & (70 - 3)(0.60 - 0.50)^2 + (65 - 3)(0.70 - 0.50)^2 \\ = {} & 10.4\end{aligned}$$

The probability associated with this χ^2 based on $(k - 1)$ 9 degrees of freedom is $p = .3182$. Thus, the effect sizes of the 10 hypothesis tests reported in Table 7.1 are not significantly different from one another.

DISJOINT CLUSTER ANALYSIS OF EFFECT SIZES

Hedges and Olkin (1983) described a disjoint cluster analysis procedure that goes a little bit beyond the direct approach to variability represented by the diffuse comparison procedures. This procedure identifies "disjoint" or nonoverlapping clusters of hypothesis tests, based on the effect size (r and Z_{FISHER}) and the sample size (N) associated with each hypothesis test (cf. Gibbons, Olkin, & Sobel, 1977, for a thorough discussion of clustering techniques at the primary level of analysis). In other words, this procedure identifies distinct clusters or groups of hypothesis tests that significantly differ from one another.

From the statistical test of the hypothesis reported for each study, a product-moment r is obtained for effect size, and Z_{FISHER}s are in turn derived from these rs. Next, these Z_{FISHER}s are rank ordered from highest to lowest. Each of these rank ordered Z_{FISHER}s is then weighted by a square-mean-root estimate of common sample size to yield a standard normal variate, U:

$$U = \left(\frac{\Sigma\sqrt{N_j - 3}}{k}\right) Z_{\text{FISHER}_j}$$

where: N_j = N associated with hypothesis test j

Z_{FISHER} = Z_{FISHER} associated with effect size for hypothesis test j

k = number of hypothesis tests included in meta-analysis

The difference between adjacently ranked *U*s is then examined. If two adjacently ranked *U*s are significantly different, then the corresponding two hypothesis tests fall into different clusters; the significant difference between two adjacently ranked *U*s defines the border between one disjoint cluster and the next.

According to this approach, significance of the difference between adjacently ranked *U*s is dependent on the number of hypothesis tests included in the meta-analysis. Hedges and Olkin (1983, Table 1) provided a table of critical values for determining whether a difference between adjacently ranked effect sizes is significant, given the number of hypothesis tests involved and given a specified level of statistical significance. If the difference between two adjacently ranked *U*s is equal to or exceeds the critical value for a selected level of statistical significance, the corresponding two hypothesis tests are placed in different clusters; otherwise, these two hypothesis tests are placed in the same cluster.

For example, consider the calculation of the disjoint cluster analysis for the $k = 10$ hypothesis tests presented in Table 7.1. First, the square-mean-root estimate of common sample size is derived as:

$$\frac{\Sigma\sqrt{N_j - 3}}{k} = \frac{\sqrt{(110-3)} + \sqrt{(100-3)} + \sqrt{(90-3)} + \sqrt{(80-3)} + \sqrt{(80-3)} + \sqrt{(70-3)} + \sqrt{(60-3)} + \sqrt{(65-3)} + \sqrt{(70-3)} + \sqrt{(65-3)}}{10}$$

$$= 8.71$$

Then, the standard normal variate, *U,* can be computed for each adjacently ranked hypothesis test by multiplying this square-mean-root estimate of common sample size by the corresponding ranked Z_{FISHER}s for effect size:

10. U1 = (8.71) (0.70) = 6.097
2. U2 = (8.71) (0.62) = 5.400
9. U3 = (8.71) (0.60) = 5.226
5. U4 = (8.71) (0.59) = 5.139
6. U5 = (8.71) (0.50) = 4.355
1. U6 = (8.71) (0.49) = 4.268
3. U7 = (8.71) (0.40) = 3.484
8. U8 = (8.71) (0.39) = 3.397
4. U9 = (8.71) (0.35) = 3.049
7. U10 = (8.71) (0.31) = 2.700

Then, the difference between adjacently ranked *U*s is obtained and examined:

U1 - U2 = 6.097 − 5.400 = 0.697
U2 - U3 = 5.400 − 5.226 = 0.174
U3 - U4 = 5.226 − 5.139 = 0.087
U4 - U5 = 5.139 − 4.355 = 0.784
U5–U6 = 4.355 − 4.268 = 0.087
U6–U7 = 4.268 − 3.484 = 0.784
U7–U8 = 3.484 − 3.397 = 0.087
U8–U9 = 3.397 − 3.049 = 0.348
U9–U10 = 3.049 − 2.700 = 0.349

The critical value for a set of $k = 10$ hypothesis tests at the significance level of $p = .05$ is 1.76 (Hedges & Olkin, 1983, Table 1). None of the differences between adjacently ranked clusters are equal to or greater than this critical value; hence, the effect sizes of the 10 hypothesis tests presented in Table 7.1 fall into a single "cluster" at the $p = .05$ level of significance.

These results are consistent with the nonsignificant diffuse comparison of effect sizes reported earlier in this chapter. The disjoint cluster analysis revealed that these 10 hypothesis tests fall into a single cluster at the $p = .05$ level of significance, and the diffuse comparison of effect sizes indicated that the effect sizes of these 10 hypothesis tests are not significantly different from one another.

Disjoint cluster analysis provides an interesting approach to variability in study outcomes, which may also set the stage for the third general analytic question of prediction and explanation. Once disjoint clusters have been identified, the reviewer may be able to work "backwards" from these clusters to some underlying explanation. This might be done by rereading the procedure and methods sections of the included studies while looking for similarities within clusters and differences between clusters. The clusters of hypothesis tests may be found a posteriori to differ on some previously unidentified study attribute. This study attribute may then be used to begin the focused comparison approach to prediction and explanation, described in the next chapter. Thus, the disjoint cluster analysis approach might be thought of, not unkindly, as a sophisticated and elegant "fishing expedition," which can help the careful reviewer to uncover previously hidden moderators of study outcomes. As with any unplanned contrast, the significance levels associated with post hoc comparisons suggested by this type of approach should be conservatively interpreted (Rosenthal & Rosnow, 1984).

FREQUENCY DISTRIBUTIONS

Finally, a word should be said on behalf of frequency distributions. Light and Pillemer (1984), who introduced the funnel plot into the meta-analytic arena, have probably emphasized the value of the visual inspection of graphic presentations of meta-analytic information more than any other authors in meta-analysis. In this instance, the visual inspection of the frequency distribution has much to recommend it. For example, the law of large numbers and the central limit theorem suggest that a reasonably large number of hypothesis tests, each of which is based on a reasonably large sample size, should produce a distribution of effect sizes which are relatively normally distributed. This means that the distribution of effect sizes should take the form of a neat, symmetrical bell-shaped curve. Visual inspection of the frequency distribution of effect sizes can help to identify the specific pattern represented by a significant diffuse comparison of study outcomes: Is the distribution bimodal (i.e., with two modes, or two different "typical" study outcomes)? Is the distribution skewed (i.e., with many of the observations trailing asymmetrically off to one side of the average)? Or, is the distribution simply spread out, symmetrically, across a wide range of study outcomes?

For example, Fig. 7.1 presents the frequency distribution for the k = 10 hypothesis tests presented in Table 7.1.

This meta-analytic database is too small to dramatically illustrate a skewed or bimodal distribution. What becomes apparent from this particular frequency distribution is that there is a considerable range of effect sizes (from Z_{FISHER} = 0.31 to Z_{FISHER} = 0.70), and that the modal study outcome (approximately Z_{FISHER} = 0.60 [actually representing the Z_{FISHER} = 0.59 for hypothesis test 5, and Z_{FISHER} = 0.60 for hypothesis test 9]) is quite similar, but not identical, to the weighted mean of Z_{FISHER} = 0.499. In addition, this frequency distribution illustrates the pattern producing the results of the disjoint cluster analysis reported before this. That is, there are no dramatic gaps between groups or clusters of studies in this frequency distribution.

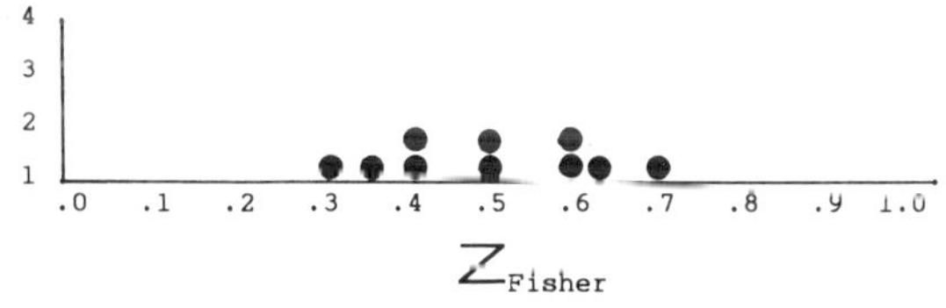

FIG. 7.1. Frequency distribution of effect sizes presented in Table 7.1.

As Light and Pillemer (1984) noted:

> These simple procedures for constructing distributions are descriptive rather than inferential. They do not provide a definitive "test" of the population distribution. But a graph can present a warning signal when outcomes are clearly incongruent with a one-population-value situation. (p. 61)

Thus, graphical displays like the frequency distribution can help the reviewer to identify important aspects of the variability around the typical study outcome. However, their most effective use is in conjunction with precise meta-analytic statistical procedures that can be used to gauge the significance of these visual patterns.

8

Focused Comparisons of Significance Levels and Effect Sizes (Simple Effects)

Science is built up with facts, as a house is with stones. But a collection of facts is no more a science than a heap of stones is a house.

—Jules Henri Poincare (1908)

The primary-level researcher begins by asking, "What is the typical response," or "What is the average," and progresses to the question of "How much variability is there around that typical response?" Eventually, however, the primary-level researcher moves on to the general analytic question of prediction and explanation in an effort to account for and make sense of that variability around the typical result. It is only with an eye toward prediction and explanation that the primary-level researcher begins to address the fundamental question of "why?"

Similarly, our overview of meta-analytic techniques has so far considered the combinations of study outcomes, which renders a gauge of the typical, combined effects obtained in a research domain. In addition, we have considered the diffuse comparisons of study outcomes that render a gauge of the amount of variability around that typical study outcome. However, it is only with an eye toward prediction and explanation that the meta-analytic reviewer can come to understand the causes of and contributions to the phenomenon represented in the well-specified hypothesis test. Our emphasis in this chapter is on the general analytic question of prediction and explanation: To what extent do the significance levels and effect sizes vary in meaningful, predictable ways as a function of practically important or theoretically illuminating variables?

Recall that four general sources of predictors of study outcomes were delineated in chapter 5: direct codings, judges' ratings, post hoc theoretical indices, and predictors derived from archival and historical sources. These various types of predictors can be used to perform the type of comparison identified as a *focused comparison* of significance levels or effect sizes. Once again, as identified in the previous chapter, these types of comparisons are focused in the sense of focusing on some specific, directional, well-defined account for how study outcomes differ. In this chapter, we consider some of the simpler, more fundamental types of focused comparisons of study outcomes, and in the next chapter we address some of the more complex and elaborate types of focused comparisons of study outcomes.

Once again, we employ a given set of 10 hypothesis tests to illustrate the calculation and interpretation of these meta-analytic statistics. The $k = 10$ hypothesis tests used before, along with the corresponding statistical information and predictors, are presented once again in Table 8.1.

Note, once again, that the statistics derived for each hypothesis test have already been transformed into the common metrics of Z and one-tailed p (for significance level), and Z_{FISHER}, r, r^2, and d (for effect size). These transformations are obtained using the algorithms presented in chapter 4. Just as in the previous sets of analyses looking at combinations and diffuse comparisons of study outcomes, we perform our computations for focused comparisons on the two sets of Zs: the standard normal deviate Z for significance level, and the Fisher's Z transformation of r for effect size.

CONTINUOUS PREDICTORS: CORRELATIONS BETWEEN PREDICTORS AND STUDY OUTCOMES

We begin by considering continuous types of predictors; that is, predictors that can be represented by values anywhere along some scale of numbers. Examples of continuous predictors, which we consider later in this chapter, are year of publication and the magnitude of the manipulation of the independent variable.

A first step in trying to gauge the extent to which study outcomes can be accounted for by a particular predictor is to examine the correlation between study outcomes and that predictor. As discussed in chapter 1, the application of statistical techniques developed for the primary level of analysis to meta-analytic data points is highly problematic; the

Table 8.1
Hypothesis Tests Included in Meta-Analytic Data Base

Study	Statistic (*df*)	*N*	DOE	Significance *Z*	*p*	Effect Size Z_{FISHER}	r	r^2	d	Year	Mag. of *X*	*OP*
1	$\chi^2 = 23(1)$	110	+	4.80	8.35E − 7	0.49	.457	.209	1.03	1981	7	0
2	$t = 6.5(98)$	100	+	5.91	2.08E − 9	0.62	.549	.301	1.31	1984	5	0
3	$F = 15.0(1,88)$	90	+	3.71	.00010	0.40	.382	.146	0.83	1985	8	1
4	$r = .335(78)$	80	+	3.04	.00119	0.35	.335	.112	0.71	1982	5	0
5	$p = .000001(-)$	80	+	4.75	.000001	0.59	.531	.282	1.25	1983	8	0
6	$Z = 3.891(-)$	70	+	3.891	.00005	0.50	.465	.216	1.05	1988	3	1
7	$p = .01(-)$	60	+	2.33	.01	0.31	.300	.090	0.63	1990	2	1
8	$F = 10.25(1,53)$	65	+	3.07	.00107	0.39	.374	.140	0.81	1986	4	0
9	$r = .535(68)$	70	+	4.77	9.45E − 7	0.60	.535	.286	1.27	1987	10	1
10	$t = 6.0(63)$	65	+	5.31	5.79E − 8	0.70	.603	.364	1.51	1989	9	1

Note: DOE = Direction Of Effect (where + = expected direction,
− = unexpected direction)
Mag. of *X* = Magnitude of manipulation of independent variable *X*
OP = Observer Presence (where 1 = Present,
0 = Absent)

assumptions required for the defensible application of primary-level inferential statistical techniques are likely to be violated by the meta-analytic data points of significance level and effect size. What this means, in the context of examining the correlation between predictors and study outcomes, is that the correlation between predictors and study outcomes can be derived and examined as a preliminary gauge of the predictive power of that particular predictor, but that the significance of that prediction is not appropriately determined by conducting the standard inferential statistical test of the significance of that correlation coefficient (cf. Cooper, 1984; R. Rosenthal, 1984, for further discussions). Still, the correlation between study outcomes and predictors is an illuminating and informative index of the extent to which study outcomes might be accounted for by the predictor under consideration.

There are two conventions about the use of correlations in meta-analyses that should be made explicit. First, correlations between predictors and indices of effect size are examined more often than correlations between predictors and indices of significance level. The assumption underlying this convention seems to be that (in light of the relations between significance level, effect size, and sample size delineated in chapter 4) effect size is somehow the more informative study outcome, and that significance level is merely a reflection of the effect size in conjunction with sample size. There is nothing wrong with examining the correlation between predictors and indices of significance levels; however, in the present context, we follow the convention of examining correlations between predictors and effect size.

Second, not all three of the primary indicators of effect size (i.e., r, Z_{FISHER}, and d) are used equally often in correlational analyses involving predictors. These three indicators of effect size are themselves highly correlated. For example, for the $k = 10$ hypothesis tests presented in Table 8.1, the correlation between Z_{FISHER} and r is $r = .9797$; the correlation between Z_{FISHER} and d is $r = .9998$; and, the correlation between r and d is $r = .9814$. However, once again, the distribution of r tends to become skewed at the extreme end of its metric. Therefore, it is often the case that the metrics of Z_{FISHER} and d are preferred to that of r in correlating predictors with study outcomes. Although the correlations between predictors and r for effect size can certainly be examined, for the sake of illustration we use Fisher's Z transformation of r as the indicator of effect size to be correlated with predictors.

For example, consider the prediction of Z for effect size by the two "continuous" study attributes represented in Table 8.1: year of publication and magnitude of the manipulation of the independent variable X.

For year of publication, the correlation is $r = .040$. Although testing the significance of this correlation is, once again, not the appropriate way to gauge the significance of the prediction at the meta-analytic level, we can note that the level of significance associated with a correlation of this magnitude based on 10 data points is $p = .4568$. The correlation between year of publication and Z_{FISHER} is depicted in Fig. 8.1.

Alternatively, for magnitude of the manipulation of *X*, the correlation is $r = .629$. The level of significance associated with a correlation of this magnitude based on 10 data points is $p = .0261$ (although, again, this is not the appropriate statistical test of the significance of this meta-analytic prediction). The correlation between magnitude of manipulation of *X* and Z_{FISHER} is depicted in Fig. 8.2..

Thus, these correlations provide a useful first step in gauging extent to which study outcomes can be predicted by some theoretically relevant or practically important characteristics.

CONTINUOUS PREDICTORS: FOCUSED COMPARISON OF SIGNIFICANCE LEVELS

This meta-analytic procedure allows the meta-analyst to estimate the probability that the *p* values of the *k* included hypothesis tests are significantly predictable in some systematic, specifiable manner. In other words, a significant focused comparison of significance levels indicates that the probability levels of the included hypothesis tests covaried significantly with the predictor being examined in that particular focused comparison. From the statistical test of the hypothesis reported for each

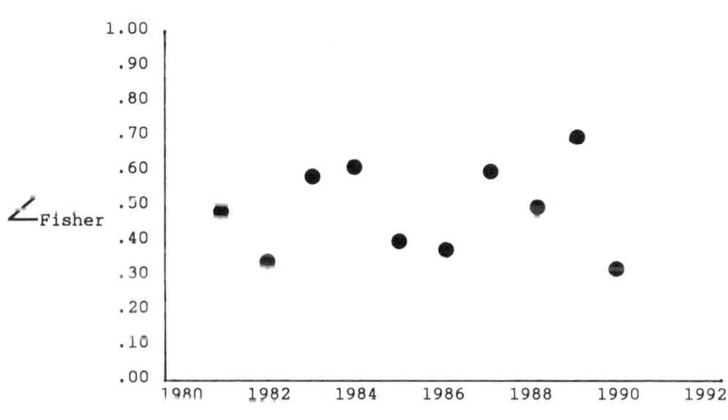

FIG.8.1. Effect size and year of publication.

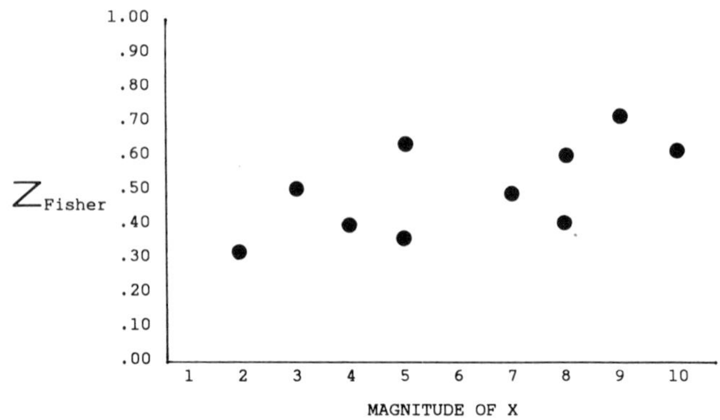

FIG.8.2. Effect size and magnitude of manipulation of *X*.

study, a *Z* is obtained for significance level. Then, each *Z* is multiplied by a contrast weight, λ. These weighted *Z*s are then integrated, using a formula that is analogous to that used in the simple combination of significance levels described in chapter 6:

$$Z = \frac{\Sigma\lambda_j Z_j}{\sqrt{\Sigma\lambda_j^2}}$$

where: λ_j = contrast weight assigned to results of hypothesis test *j*
Z_j = *Z* associated with significance level of hypothesis test *j*

The *p* value corresponding to this *Z* for focused comparison is the likelihood that the *k* study outcomes varied as a function of the predictor represented by the included λs.

We have not yet addressed these contrast weights, λs, and we should take a moment to consider them. These λs generally represent linear orthogonal polynomial contrast weights (identical to those used in chapter 2 to derive a more focused *F* test comparing cell means in the context of primary-level studies). For example, if there are only three hypothesis tests included in the meta-analytic data base, and these three hypothesis tests were published in 1980, 1981, and 1982, then the λs for these three studies representing year of publication would be: $\lambda_{1980} = -1$; $\lambda_{1981} = 0$; and, $\lambda_{1982} = +1$. These contrast weights capture the variation between hypothesis tests on the predictor under consideration, with the constraint that the sum of these λs must equal zero. For neat and even cases (such as the three year-of-publication λs just presented), one can derive the relevant λs from tables in the back of many statistics books. However, in most meta-analyses, there will be unequal intervals between levels of the predictor under consideration, and/or unequal numbers of exemplars at each level of the predictor. In this more common circumstance, the procedure described in Kirk (1982, Appendix C) derives contrast weights that take into account unequal intervals and unequal numbers of exemplars, and still sum to equal zero.

For example, Table 8.2 presents the year and magnitude of X for each of the $k = 10$ hypothesis tests presented in Table 8.1, along with the derived λs.

The Z for focused comparison of significance levels for year of publication is derived as:

$$Z = \frac{-9(4.80) - 3(5.91) - 1(3.71) - 7(3.04) - 5(4.75) + 5(3.891) + 9(2.33) + 1(3.07) + 3(4.77) + 7(5.31)}{\sqrt{(-9)^2 + (-3)^2 + (-1)^2 + (-7)^2 + (-5)^2 + (5)^2 + (9)^2 + (1)^2 + (3)^2 + (7)^2}}$$

$$Z = 0.807$$

The probability associated with this Z for focused comparison for year of publication is $p = .2099$. Thus, hypothesis tests in this meta-analytic database do not seem to yield more significant results in more recent/earlier studies.

The Z for focused comparison of significance levels for magnitude of manipulation of X is derived as:

$$Z = \frac{9(4.80) - 11(5.91) + 19(3.71) - 11(3.04) + 19(4.75) - 31(3.891) - 41(2.33) - 21(3.07) + 39(4.77) + 29(5.31)}{\sqrt{(9)^2 + (-11)^2 + (19)^2 + (-11)^2 + (19)^2 + (-31)^2 + (-41)^2 + (-21)^2 + (39)^2 + (29)^2}}$$

$$Z = 2.051$$

Table 8.2
Predictors and λs for 10 Hypothesis Tests

Study	*Year*	*Year λ*	*Mag. X*	*Mag.Xλ*
1	1981	−9	7	+9
2	1984	−3	5	−11
3	1985	−1	8	+19
4	1982	−7	5	−11
5	1983	−5	8	+19
6	1988	+5	3	−31
7	1990	+9	2	−41
8	1986	+1	4	−21
9	1987	+3	10	+39
10	1989	+7	9	+29

The probability associated with the focused comparison of significance levels for magnitude of manipulation of X is $p = .0201$. Thus, hypothesis tests in this meta-analytic database do yield more significant results if stronger manipulations of X are utilized.

CONTINUOUS PREDICTORS: FOCUSED COMPARISON OF EFFECT SIZES

This meta-analytic procedure allows the meta-analyst to estimate the probability that the effect sizes of the k included hypothesis tests are significantly predictable in some systematic, specifiable manner. In other words, a significant focused comparison of effect sizes indicates that the magnitudes of effect of the included hypothesis tests covaried significantly with the predictor being examined in that particular focused comparison. From the statistical test of the hypothesis reported for each study, a product moment r is obtained for effect size, and Z_{FISHER}s are, in turn, derived from these rs. Next, each Z_{FISHER} is multiplied by a contrast weight, λ. Then, these weighted Z_{FISHER}s are integrated using a formula that is similar to the focused comparison of significance levels, which is analogous to that used in the simple combination of effect sizes described in chapter 6:

$$Z = \frac{\Sigma \lambda_j Z_{FISHER_j}}{\sqrt{\sum \frac{(\lambda_j^2)}{(N_j - 3)}}}$$

where: λ_j = contrast weight assigned to results of hypothesis test j
Z_{FISHER_j} = Fisher's Z associated with effect size of hypothesis test j
N_j = sample size associated with hypothesis test j

The probability associated with this Z for focused comparison is the likelihood that the k study outcomes varied as a function of the predictor represented by the included λs.

For example, the Z for focused comparison of effect sizes for year of publication is derived as:

$$Z = \frac{-9(0.49) - 3(0.62) - 1(0.40) - 7(0.35) - 5(0.59) + 5(0.50) + 9(0.31) + 1(0.39) + 3(0.60) + 7(0.70)}{\sqrt{\frac{(-9)^2}{107} + \frac{(-3)^2}{97} + \frac{(-1)^2}{87} + \frac{(-7)^2}{77} + \frac{(-5)^2}{77} + \frac{(5)^2}{67} + \frac{(9)^2}{57} + \frac{(1)^2}{62} + \frac{(3)^2}{67} + \frac{(7)^2}{62}}}$$

$$Z = 0.132$$

The probability associated with this Z for focused comparison for year of publication is $p = .4476$. Thus, hypothesis tests in this meta-analytic database do not seem to yield stronger effects in more recent/earlier studies.

The Z for focused comparison of effect sizes for magnitude of manipulation of X is derived as:

$$Z = \frac{9(0.49) - 11(0.62) + 19(0.40) - 11(0.35) + 19(0.59) - 31(0.50) - 41(0.31) - 21\ 1(0.39) + 39(0.60) + 29(0.70)}{\sqrt{\frac{(9)^2}{107} + \frac{(-11)^2}{97} + \frac{(19)^2}{87} + \frac{(-11)^2}{77} + \frac{(19)^2}{77} + \frac{(-31)^2}{67} + \frac{(-41)^2}{57} + \frac{(-21)^2}{62} + \frac{(39)^2}{67} + \frac{(29)^2}{62}}}$$

$$Z = 1.970$$

The probability associated with the focused comparison of effect sizes for magnitude of manipulation of X is $p = .0244$. Thus, hypothesis tests in this meta-analytic database do yield stronger effects if stronger manipulations of X are utilized. Given the theoretical context of this particular research domain, whatever X and Y might represent, it may stand to reason that stronger and more significant effects are obtained with stronger manipulations of X. Because, supposedly, the independent variable of X is supposed to cause changes in the dependent variable Y, it should not be surprising that more of X results in more of a change in Y.

CATEGORICAL PREDICTORS: CORRELATIONS, AND FOCUSED COMPARISONS OF SIGNIFICANCE LEVELS AND EFFECT SIZES

The examples we have considered up to this point have explicitly involved continuous predictors. We should recognize that the same general approach can be applied to categorical predictors; that is, predictors that can be represented by some limited and discrete set of alternatives. In the data presented in our meta-analytic database, the presence of the observer constitutes a simple, standard categorical predictor; the observer can be identified as either present (represented by a "1") or absent (represented by a "0").

As a first step, just as with continuous predictors, one can examine the correlation between study outcome and the categorical predictor. The calculation of a product-moment correlation when one variable (effect size) is on a continuous metric and the other variable (the categorical predictor) is on a dichotomous metric produces what is called a *point-biserial* correlation coefficient. For example, the correlation between observer presence and Z_{FISHER} for effect size is $r = .054$. The significance level associated with an r of this magnitude based on 10 data points is $p = .4412$ (although, once again, the significance of this relationship is not appropriately tested with this primary-level inferential statistic). The focused comparison for significance levels for observer presence is derived using the same formula as the one presented earlier:

$$Z = \frac{-1(4.80) - 1(5.91) + 1(3.71) - 1(3.04) - 1(4.75) + 1(3.891) + 1(2.33) - 1(3.07) + 1(4.77) + 1(5.31)}{\sqrt{(-1)^2 + (-1)^2 + (1)^2 + (-1)^2 + (-1)^2 + (1)^2 + (1)^2 + (-1)^2 + (1)^2 + (1)^2}}$$

$$Z = 0.491$$

The probability associated with this Z for focused comparison of significance levels is $p = .3117$. Similarly, the focused comparison for effect sizes for observer presence is derived using the same formula as presented earlier:

$$Z = \frac{-1(0.49) - 1(0.62) + 1(0.40) - 1(0.35) - 1(0.59) + 1(0.50) + 1(0.31) - 1(0.39) + 1(0.60) + 1(0.70)}{\sqrt{\frac{(-1)^2}{107} + \frac{(-1)^2}{97} + \frac{(1)^2}{87} + \frac{(-1)^2}{77} + \frac{(-1)^2}{77} + \frac{(1)^2}{67} + \frac{(1)^2}{57} + \frac{(-1)^2}{62} + \frac{(1)^2}{67} + \frac{(1)^2}{62}}}$$

$$Z = 0.179$$

The probability associated with this Z for focused comparison of effect sizes is $p = .4288$. Thus, neither the significance levels nor the effect sizes of the 10 hypothesis tests reported in Table 8.1 appear to be influenced by the presence or absence of the observer.

Note that these focused comparisons for the categorical variable use λs of +1 and −1 in this instance. However, an unequal number of exemplars in each of the two categories would have required different contrast weights. For example, if three of the hypothesis tests were conducted with the observer present, and five were conducted with the observer absent, then the a focused comparison that contrasted these two categories of studies would use a $\lambda = +5$ for the 3 "present" studies, and a $\lambda = -3$ for the 5 "absent" studies. As always, these contrast weights must (a) capture the variation in the predictor of interest, and, (b) sum to equal zero.

BLOCKING: ANOTHER ILLUSTRATION OF CATEGORICAL PREDICTORS

One way of illustrating the effects of a categorical predictor (described previously) is with the point-biserial correlation produced by correlation effect size with the dicotomous predictor. Although this can be informative, it does not completely convey the difference between the typical study outcomes at the two levels of a dichotomous categorical predictor. In addition, three or more levels of a categorical predictor that are *not* arranged on any ordinal metric will not yield meaningful correlations between categorical predictor level and effect sizes.

An alternative illustration of the effects of a categorical predictor can be produced through the technique of classification, or *blocking*. This simply requires that the k hypothesis tests be blocked, or classified into separate groups, based on the levels of the categorical predictor under consideration. Then, combinations of significance levels and effect sizes, as presented in the previous chapter, can be derived within each level of the categorical predictor.

For example, suppose we were to block the $k = 10$ hypothesis tests presented in Table 8.1 into those hypothesis tests where the observer was absent ($k = 5$) and those where the observer was present ($k = 5$). The combinations of significance levels and effect sizes for the observer-absent hypothesis tests yields $Z = 9.815$, $p = 2.12\text{E}-20$, $\overline{Z}_{\text{FISHER}} = 0.498$, $\bar{r} = .461$, $\bar{r}^2 = .212$, $\bar{d} = 1.039$. The combinations for the observer-present hypothesis tests yields $Z = 8.887$, $p = 1.07\text{E}-17$,

$\overline{Z}_{\text{FISHER}} = 0.499$, $\bar{r} = .461$, $\bar{r}^2 = .213$, $\bar{d} = 1.040$. This illustrates the pattern that produced the nonsignificant focused comparisons of significance levels and effect sizes as a function of observer presence.

A number of general issues regarding the focused comparisons of simple effects should now be considered.

CONTINUOUS PREDICTORS VERSUS CATEGORICAL PREDICTORS

Recall from the primary-level illustration presented at the end of chapter 4 that categorical treatments of continuous data have the potential to lose informative variability and yield weaker tests of the hypothesis under consideration. For example, a correlational analysis of the data presented in Table 4.3 resulted in an effect size of $r = .9503$, whereas a t test of that same data (collapsing one of the continuous metrics into a dichotomous categorical variable) resulted in an effect size of $r = .842$.

The same sort of distortion can occur at a meta-analytic level when an inherently continuous predictor is treated as a categorical predictor. For example, consider the $k = 10$ hypothesis tests presented in Table 8.3. This table is identical to Table 8.1 except for the treatment of the strength of the manipulation of the independent variable X. Any hypothesis test that used a magnitude of $X < 6$ has been identified in Table 8.3 as having a "type of X" equal to zero (i.e., $0 =$ low X). Any hypothesis test that used a magnitude of $X > 6$ has been identified in Table 8.3 as having a type of X equal to one (i.e., $1 =$ high X).

When we treat this inherently continuous predictor as a dichotomous predictor, we find a considerable dissipation of this variable's predictive power. For example, the correlation between this categorical predictor and effect size is $r = .498$ (compared to the $r = .629$ obtained when this variable was treated as a continuous predictor). The focused comparisons of significance levels and effect sizes for this categorical predictor ($Z = 1.616$, $p = .0531$; $Z = 1.652$, $p = .0493$) are noticeably weaker than those obtained from the continuous predictor ($Z = 2.051$, $p = .0201$; $Z = 1.970$, $p = .0244$). Thus, in order to take full advantage of the predictive power of a predictor, variables on continuous metrics should be left on continuous metrics whenever at all possible.

CONTINUOUS PREDICTORS FROM CATEGORICAL PREDICTORS

Following from the previous point, it is interesting to consider instances in which it appears that only a categorical predictor can be coded for each hypothesis test included in the database, and yet which might be used

Table 8.3
Hypothesis Tests Included in Meta-Analytic Data Base

Study	Statistic (df)	N	DOE	Significance Effect Size						Year	Type of X	OP
				Z	p	Z_{FISHER}	r	r^2	d			
1	$\chi^2 = 23(1)$	110	+	4.80	8.35E-7	0.49	.457	.209	1.03	1981	1	0
2	$t = 6.5(98)$	100	+	5.91	2.08E-9	0.62	.549	.301	1.31	1984	0	0
3	$F = 15.0(1,88)$	90	+	3.71	.00010	0.40	.382	.146	0.83	1985	1	1
4	$r = .335(78)$	80	+	3.04	.00119	0.35	.335	.112	0.71	1982	0	0
5	$p = .000001(-)$	80	+	4.75	.000001	0.59	.531	.282	1.25	1983	1	0
6	$Z = 3.891(-)$	70	+	3.891	.00005	0.50	.465	.216	1.05	1988	0	1
7	$p = .01(-)$	60	+	2.33	.01	0.31	.300	.090	0.63	1990	0	1
8	$F = 10.25(1,63)$	65	+	3.07	.00107	0.39	.374	.140	0.81	1986	0	0
9	$r = .535(68)$	70	+	4.77	9.45E-7	0.60	.535	.286	1.27	1987	1	1
10	$t = 6.0(63)$	65	+	5.31	5.79E-8	0.70	.603	.364	1.51	1989	1	1

Note: DOE = Direction Of Effect (where + = expected direction,
− = unexpected direction)
Type of X = Magnitude of manipulation of independent variable X
(where 1 = high X, 0 + low X)
OP = Observer Presence (where 1 = Present,
0 = Absent)

to derive potentially more informative continuous predictors. If each hypothesis test compares the mean *Y* obtained under conditions of high *X* and low *X*, this clearly constitutes the categorical manipulation of the independent variable *X*. However, if the specific values of *X* for the high and low conditions are reported for each hypothesis test (e.g., the number of minutes exposure for high vs. low exposure; the amount of dosage per unit body weight for high vs. low dosage), a simple ratio can be developed to help reconstruct a continuous metric of *X* across hypothesis tests.

Specifically, for each hypothesis test, a new predictor could be defined as (the value of the high *X*) / (the value of the low *X*). Then, after correlating effect size with this ratio predictor, the reviewer can reconstruct a picture of the relation between effect size and *X* (now defined as a continuous predictor). This is done by determining the predicted effect size at each level of "high *X*" where "low *X*" is set equal to 1. This yields a continuous metric of *X* that ranges from 1 to the most extreme value of *X* included in the data base.

For example, Glass (1980) developed this approach in the estimation of a model of the effects of class size on achievement. One study in this particular research domain may have compared the achievement of students in a large class (class size of 50) with the achievement of students in a small class (class size of 25). Another study may have compared the achievement of students in a large class (class size of 25) with the achievement of students in a small class (class size of 5). Note that what constituted a small class (low *X*) in one study constituted a large class (high *X*) in the other study. By performing a regression analysis (Kerlinger & Pedhauzur, 1973), where the effect size for each hypothesis test was regressed upon the ratio of the size of the larger class divided by the size of the smaller class, Glass demonstrated that the achievement decreased as a function of class size.

Note that the value of this approach does *not* lie in testing the significance of this new, continuous predictor in predicting study outcomes. The significance of this new continuous predictor would be gauged by the appropriate focused comparisons of significance levels and effect sizes. Rather, this strategy simply allows the reviewer to put a series of categorical manipulations of the predictor back on to their implicit continuous metric.

PRIMARY-LEVEL STATISTICS VERSUS META-ANALYTIC-FOCUSED COMPARISONS

We have considered the use of correlations between study outcomes and predictors as an illuminating first step in trying to answer the general analytic question of prediction and explanation. At the same time, we have repeatedly emphasized that the significance of the predictive power of a particular predictor is not appropriately determined by conducting the standard inferential statistical test of the significance of that correlation coefficient. Again, the reason for this is that the assumptions required for the defensible application of the primary-level inferential statistical techniques may be violated by the meta-analytic data points of significance level and effect size (cf. Cooper, 1984; Hedges & Olkin, 1985; R. Rosenthal, 1984; and Strube & Hartman, 1983, for discussions).

The attentive reader will have noticed that the results of testing the significance of the correlation between predictors and study outcomes reported earlier generally support the results of the corresponding meta-analytic focused comparisons. For example, the correlation between year of publication and Z_{FISHER} for effect size was $r(8) = .040$, a correlation associated with a probability level of $p = .4568$; the corresponding meta-analytic focused comparison of effect sizes yielded a $Z = 0.132$, $p = .4476$. Similarly, the correlation between magnitude of manipulation of X and Z_{FISHER} for effect size was $r(8) = .629$, a correlation associated with a probability level of $p = .0261$; the corresponding meta-analytic focused comparison of effect sizes yielded a $Z = 1.970$, $p = .0244$. Thus, the application of primary-level inferential statistics can sometimes support the results of meta-analytic focused comparisons.

However, the reader should be cautious of a logical fallacy that may be hidden here: Just because some university professors are male, it does not follow that *all* university professors are male. By the same token, just because some applications of primary-level inferential statistics to meta-analytic data points may support the results of meta-analytic focused comparisons, it does not follow that *all* such applications will support the results of meta-analytic focused comparisons.

Consider one example: Mullen and Riordan (1988) reported the results of meta-analytic integrations of the self-serving attribution hypothesis in sports settings for several different types of measures. The self-serving attribution hypothesis suggests that people will be more likely accept responsibility for success, and less likely to accept the blame for failure. One particular meta-analytic integration examined this phenomenon for hypothesis tests using a general internal–external scale of attribution. The $k = 7$ hypothesis tests that used this specific measure produced significant and moderate-sized effects, with a significant degree of heterogeneity of both significance levels and effect sizes. However, of particular interest in this context is the prediction of these effects by the size of the team. The correlation coefficient between team size and effect size was not significant using the primary-level inferential statistical techniques: for Cohen's d, $r(5) = .550$, $p = .1015$; for Z_{FISHER}, $r(5) = .527$, $p = .1131$. However, the meta-analytic focused comparisons for significance levels, $Z = 2.672$, $p = .0038$, and effect sizes, $Z = 2.335$, $p = .0098$, were both highly significant (cf. Mullen & Riordan, 1988, for a discussion of the theoretical implications of this particular pattern).

The literature provides a number of similar examples of the failure of primary-level inferential statistics, applied to meta-analytic data points, to support the results of meta-analytic focused comparisons (e.g., Mullen & Felleman, in press; Mullen & Hu, 1988). The point is not that correlations between study outcomes and theoretically relevant predictors are wrong; these correlations provide a quick and intelligible first glance at the capacity of a predictor to account for study outcomes. However, the significance of each predictive relationship should be gauged by the appropriate meta-analytic focused comparison.

DIFFUSE COMPARISONS AND FOCUSED COMPARISONS

An interesting comparison between the value, and the results, of diffuse comparisons and focused comparisons is provided by the data base presented in Table 8.1. Recall, from chapter 7, that these meta-analytic data points yielded nonsignificant diffuse comparisons of significance levels and effect sizes. This indicated that the basic combinations of significance levels and effect sizes were probably rather good characterizations of the typical outcomes in this research domain, and that the hypothesis tests were not likely to have been sampled from different populations. However, a nonsignificant diffuse test does *not* mean that there is no interesting variability in the research domain. Indeed, the meta-analytic data points presented in Table 8.1, which failed to yield significant diffuse comparisons, are demonstrated with these focused comparisons to vary in systematic and predictable ways as a function of the magnitude of the manipulation of the independent variable (although not as a function of year of publication or observer presence).

The point is that diffuse comparisons and focused comparisons are designed to answer very different questions (notably, as we have seen, the general analytic questions of variability and prediction, respectively). One might find, as we have found with this meta-analytic data base, significant focused comparison predictions in the absence of significant diffuse heterogeneity. Of course, one could find significant heterogeneity with the diffuse comparisons and still be able to significantly account for variability in study outcomes with a theoretically relevant focused comparison. The other two possibilities (i.e., the failure to derive any significant focused comparisons, with either significant or nonsignificant diffuse comparisons) are entirely possible as well. These last two possibilities may be most likely either when there is a considerable amount of random variability in study outcomes (perhaps due to weak or unreliable measurements of the dependent variable), or when the meta-analytic reviewer has simply not uncovered the theoretically relevant predictors that would account for variation from one study to the next.

The questions of variability (diffuse comparisons) and prediction (focused comparisons) are as different from one another as each one is from the initial question of typical outcome (combinations). The reviewer should not be disinclined to pursue a theoretically meaningful focused comparison by a nonsignificant diffuse comparison. By the same token, the reviewer should not interpret a significant diffuse comparison as a guarantee of finding a significant focused comparison.

9

Focused Comparisons of Significance Levels and Effect Sizes (Complex Effects)

Every conclusion drawn from our observation is, as a rule, premature, for behind the phenomena which we see clearly are other phenomena that we see indistinctly, and perhaps behind these latter, yet others which we do not see at all.

—Gustave LeBon (1895)

In this chapter we extend and continue our discussion of prediction and explanation at the meta-analytic level. In most of the instances considered in this chapter, the same basic techniques of correlation, focused comparisons of significance levels and effect sizes, and blocking are implemented. However, what makes the instances considered in this chapter different is the complexity of these implementations. What if the meta-analyst is interested in looking at the combined effects of two (or more) predictors on significance level and effect size? What if those two predictors are both categorical, or are both continuous, or one is categorical and one is continuous? Beyond the testing of the effects of multiple predictors, the meta-analytic question itself might be multivariate in nature. In other words, what if the hypothesis under consideration is not simply that X influences Y, but that X influences Y through (and only through) its effect on Z?

It is unlikely that we will be able to anticipate every possible permutation of complex effects in the testing of theoretical models at the meta-analytic level. However, in this chapter we (hopefully) consider an array of complex effects that will adequately suggest to the meta-analytic reviewer alternative, and more sophisticated, approaches to prediction and explanation at the meta-analytic level.

Once again, we employ a given set of 10 hypothesis tests to illustrate the calculation and interpretation of these meta-analytic focused comparisons. The $k = 10$ hypothesis tests used before, along with the corresponding statistical information and predictors, are presented once again in Table 9.1. Note that, once again, *Type of X* replaces *Magnitude of X*, representing a dichotomous classification of the continuous predictor we have used previously (where a type of $X = 1$ indicates that the magnitude of X fell above the midpoint of 6 [5 hypothesis tests], and a type of $X = 0$ indicates that the magnitude of X fell below the midpoint of 6 [5 studies]).

Note, once again, that the statistics derived for each hypothesis test have already been transformed into the common metrics of Z and one-tailed p (for significance level), and Z_{FISHER}, r, r^2, and d (for effect size). These transformations are obtained using the algorithms presented in chapter 4. Just as in the application of previous meta-analytic techniques, we perform our computations for focused comparisons on two sets of Zs: the standard normal deviate Z for significance level, and the Fisher's Z transformation of r for effect size.

TWO CATEGORICAL PREDICTORS

Recall that, in the previous chapter, we examined each of the two categorical predictors (observer presence and type of X) separately. It was demonstrated that observer presence did not significantly predict study outcomes, whereas type of X was found to be a significant predictor of study outcomes.

However, the possibility exists that observer presence might still exert some moderating influence on study outcomes, particularly if the observed effects of another predictor (such as type of X) can be demonstrated to vary as a function of observer presence. In other words, the simple effects analyses presented in the previous chapter indicate that observer presence does not, in and of itself, produce a more extreme effect of X upon Y. However, the influence of the type of X on the effect of X upon Y may be stronger when the observer is present, and weaker when the observer is absent. There are three general strategies for examining the complex effects of two categorical predictors: Inspect the separate subsamples of hypothesis tests defined by the crossing of both categorical predictors; examine the effects of one categorical predictor

Table 9.1
Hypothesis Tests Included in Meta-Analytic Data Base

Study	Statistic (*df*)	*N*	DOE	*Significance* Z	*Significance* p	*Effect Size* Z_{FISHER}	*Effect Size* r	*Effect Size* r^2	*Effect Size* d	Year	Type of *X*	*OP*
1	$\chi^2 = 23(1)$	110	+	4.80	8.35E − 7	0.49	.457	.209	1.03	1981	1	0
2	$t = 6.5(98)$	100	+	5.91	2.08E − 9	0.62	.549	.301	1.31	1984	0	0
3	$F = 15.0(1,88)$	90	+	3.71	.00010	0.40	.382	.146	0.83	1985	1	1
4	$r = .335(78)$	80	+	3.04	.00119	0.35	.335	.112	0.71	1982	0	0
5	$p = .000001(-)$	80	+	4.75	.000001	0.59	.531	.282	1.25	1983	1	0
6	$Z = 3.891(-)$	70	+	3.891	.00005	0.50	.465	.216	1.05	1988	0	1
7	$p = .01(-)$	60	+	2.33	.01	0.31	.300	.090	0.63	1990	0	1
8	$F = 10.25(1,63)$	65	+	3.07	.00107	0.39	.374	.140	0.81	1986	0	0
9	$r = .535(68)$	70	+	4.77	9.45E − 7	0.60	.535	.286	1.27	1987	1	1
10	$t = 6.0(63)$	65	+	5.31	5.79E − 8	0.70	.603	.364	1.51	1989	1	1

Note: DOE = Direction Of Effect (where + = expected direction,
− = unexpected direction)
Type of *X* = Magnitude of manipulation of independent variable *X*
(where 1 = high *X*, 0 = low *X*)
OP = Observer Presence (where 1 = Present,
0 = Absent)

within levels of the other categorical predictor; and test for the predictive power of some arithmetic combination of the two predictors.

The first step in examining this type of combined effect of two categorical variables would be to block all of the hypothesis tests on the two categorical predictors. This *cross-classification,* or *cross-blocking,* approach in the present context yields four distinct types of hypothesis tests: tests where the observer was present and the manipulation of X was high ($k = 3$); tests where the observer was absent and the manipulation of X was high ($k = 2$); tests where the observer was present and the manipulation of X was low ($k = 2$); and, tests where the observer was absent and the manipulation of X was low ($k = 3$). Within each of these four types of tests produced by the cross-blocking of studies, a separate set of meta-analytic combinations can be conducted. The results of these separate combinations of significance levels and effect sizes are presented in Table 9.2.

A visual inspection of this table indicates that in every cell of the table, the effect of X upon Y is highly significant and of moderate-to-strong magnitude. Moreover, this cross-blocking approach reveals a slight tendency for the moderating effect of Type of X to be somewhat weaker when the observer is absent (low X tests yielding a mean Z_{FISHER} = 0.470, high X tests yielding a mean Z_{FISHER} = 0.535, a difference of 0.065) than when the observer is present (low X tests yielding a mean

Table 9.2
Combinations of Significnce Levels and Effect Sizes Resulting From Cross-Blocking on Type of X and Observer Presence

		Type of X = 0	*Type of X = 1*
Observer Absent	Z	7.198	6.674
	p	7.76E − 13	2.13E − 11
	Nfs(p = .05)	50.4	31.7
	$\bar{Z}_{FISHER}$	0.470	0.535
	$\bar{r}$	.438	.489
	$\bar{r}^2$	.192	.239
	$\bar{d}$	0.975	1.122
Observer Present	Z	4.469	7.722
	p	.000004	2.55E − 14
	Nfs(p = .05)	12.3	37.3
	$\bar{Z}_{FISHER}$	0.414	0.548
	$\bar{r}$	.392	.499
	$\bar{r}^2$	.154	.249
	$\bar{d}$	0.853	1.152

Z_{FISHER} 0.414, high X tests yielding a mean Z_{FISHER} = 0.548, a difference of .134).

The second general strategy provides a more precise gauge of this type of effect by developing focused comparisons of the theoretically interesting predictor within each level of the other predictor. These focused comparisons are obtained in exactly the same way as those presented in the previous chapter (i.e., derive orthogonal polynomial contrast weights, and combine the Zs for significance or the Z_{FISHER}s for effect sizes, using the formula presented in chapter 8). The only difference is that these focused comparisons are now conducted on smaller subsamples of hypothesis tests.

For example, focused comparisons of significance levels and effect sizes using type of X as the predictor can be conducted for the $k = 5$ hypothesis tests with the observer absent. These analyses reveal that the trend for high X tests to produce more extreme results than low X tests was nonsignificant for significance levels, $Z = 0.842$, $p = .2000$, and for effect sizes, $Z = 0.904$, $p = .1831$. Another set of focused comparisons can be conducted for the $k = 5$ hypothesis tests with the observer present. These analyses reveal that the trend for high X tests to produce more extreme results than low X tests was marginally significant for significance levels, $Z = 1.632$, $p = .0513$, and effect sizes, $Z = 1.401$, $p = .0806$. The results of these focused comparisons within a cross-blocking procedure suggest that the presence of an observer might somehow moderate the effects of type of X upon study outcomes. In other words, clarifying the patterns that were visually detected in the cross-blocking presented in Table 9.2, higher levels of X appear to lead to more extreme study outcomes, but this seems to occur most strongly in the presence of an observer. This type of pattern may be particularly illuminating at the meta-analytic level, given the research domain and the theories guiding the integration. Thus, performing focused comparisons using one categorical predictor within levels of another categorical predictor can give some indication of the extent to which the two categorical predictors may combine to influence study outcomes.

The third strategy produces an overall focused comparison, based on the entire meta-analytic data base of k hypothesis tests, which models some theoretically relevant combination of the two predictors with some arithmetic combination. For example, suppose a particular theoretical perspective predicts that the presence of the observer and the type of operationalization of X will combine in a straightforward additive manner. Conceptually, this would mean that the strongest, most significant effects should obtain with strong manipulations of X in the presence of the observer; the weakest, least significant effects should obtain with weak manipulations of X in the absence of the observer; and intermediate effects should obtain with the mixed (i.e., hi–absent and lo–present) combinations. The meta-analytic focused comparisons that test for this type of effect are functionally equivalent to the simple effect comparisons detailed in chapter 8, as well as the simple effect focused comparisons implemented within levels of some blocking variable earlier in this chapter. The difference is in the derivation of the orthogonal contrast weights, or λs. Table 9.3 illustrates the derivation of the λs for the additive combination of the two categorical predictors, type of X and observer presence.

The first step is to derive the simple effect λs for the two categorical predictors. In this instance, this is easily accomplished; equal numbers of exemplars at each level of dichotomous categorical predictors yield the simple λs of +1 and −1. These particular λs happen to be on the same metric (i.e., ranging from −1 to +1). However, this is not necessarily always the case. Therefore, in order to begin to examine any combination of the predictive power of the two categorical predictors, the corresponding λs have to be placed on the same metric. This is accomplished by converting the λs for each categorical predictor into standard normal deviates (or, Z scores) by subtracting from each the mean λ for that predictor, and dividing this difference by the standard deviation of the λs for that predictor. The mean λ is (by definition) going to equal zero (because these λs must sum to equal zero, this sum of zero divided by the k number of hypothesis tests must also equal zero). The Z scores derived

Table 9.3
Derivation of λs for Additive Combination of Two Categorical Predictors

Study	*Type of X*	λ_{TYPE}	Z_{TYPE}	OP	λ_{OP}	Z_{OP}	$Z_{TYPE} + Z_{OP}$	$\lambda_{TYPE + OP}$
1	1	+1	+.95	0	−1	−.95	0	0
2	0	−1	−.95	0	−1	−.95	−1.9	−1
3	1	+1	+.95	1	+1	+.95	+1.9	+1
4	0	1	−.95	0	−1	−.95	−1.9	−1
5	1	+1	+.95	0	−1	−.95	0	0
6	0	1	−.95	1	+1	+.95	0	0
7	0	−1	−.95	1	+1	+.95	0	0
8	0	−1	−.95	0	−1	−.95	−1.9	−1
9	1	+1	+.95	1	+1	+.95	+1.9	+1
10	1	+1	+.95	1	+1	+.95	+1.9	+1
	Mean	0			0			
	S.D.	1.054			1.054			

from the λs have the following attributes: Like the λs from which they were derived, they sum to zero; and, like the λs from which they were derived, they accurately capture the variation across hypothesis tests in the predictor being considered; however, unlike the λs, the *Z* scores for one predictor are on the same metric as the *Z* scores for another predictor. This metric of the standard normal deviate has a mean of 0 and a standard deviation of 1. Given that these *Z* scores derived from λs are on a common metric, arithmetic operations can be performed to combine them according to theoretically informative models.

For example, because we are interested in the additive combination of these two categorical predictors, we can now add together the *Z* scores derived from the λs. This produces a new column of numbers that represents the additive combination of type of *X* and observer presence. This new predictor is positive only when both of the categorical predictors were high, it is negative only when both predictors were low, and it is equal to zero only when the two predictors were mixed. Finally, these summed *Z*s are once again converted into a new array of λs, which represent the additive combination of type of *X* and observer presence.

When these new λs are used in focused comparisons of significance levels, $Z = 0.726$, $p = .2340$, and focused comparisons of effect sizes, $Z = 1.183$, $p = .1185$, there is no evidence in support of an additive combination of the effects of type of *X* and observer presence. In other words, these focused comparisons indicate that strongest effects do not occur with strong manipulations of *X* in conjunction with observer presence, and that weaker effects do not occur with weak manipulations of *X* in conjunction with observer absence.

It should be recognized that any number of possible arithmetic combinations of two (or more) predictors can be developed to model theoretically illuminating models. The meta-analyst should have a clear and precise understanding of the theoretical implications of a particular model before "fishing" around in the meta-analytic data base.

For example, multiplicative combinations are both conceptually and practically very different from additive combinations. In the next-to-last column of Table 9.3, the reviewer might have multiplied, rather than added, the *Z*s derived from the two predictors. However, this multiplicative combination represents a very different model. The additive model suggests the strongest effects for strong manipulations in the presence of an observer, and the weakest effects for weak manipulations in the absence of the observer. The multiplicative model suggests the strongest effects for strong manipulations in the presence of the observer *and* for weak manipulations in the absence of the observer, and the weakest effects for hypothesis tests conducted under the mixed conditions. This is because equally strong, positive combinations of the λ-derived *Z*s will obtain from multiplying two positive *Z*s (strong manipulation–observer presence) *and* from multiplying two negative *Z*s (weak manipulation–observer absence). Strong, negative combinations of the λ-derived *Z*s will obtain from multiplying a negative *Z* by a positive *Z* (the two mixed conditions). The theoretical implications of this type of arithmetic combination of effects for the particular research domain under consideration should be fully understood before entering into the calculation of the focused comparisons.

ONE CATEGORICAL PREDICTOR AND ONE CONTINUOUS PREDICTOR

In order to illustrate the complex effects involved with a categorical predictor and a continuous predictor, we transform the categorical predictor of type of *X* back into the continuous predictor of magnitude of *X*, from which it was distilled. The data for the $k = 10$ hypothesis tests, with the continuous predictor replaced, are presented in Table 9.4.

Recall that our consideration of the simple effects of the magnitude of *X* in the previous chapter resulted in significant focused comparisons of significance levels and effect sizes. If we are interested in the complex effects of this continuous predictor in conjunction with the categorical predictor of observer presence, there are two distinct strategies: examine the effects of the continuous predictor within each level of the categorical predictor; and, test for the predictive power of some arithmetic combination of the two predictors. A third strategy, more directly pursued previously for the complex effect of two categorical predictors, should be recognized: The continuous predictor (magnitude of *X*) could be collapsed into a categorical predictor (type of *X*), and the separate subsamples of hypothesis tests created by the crossing of these two categorical predictors can be examined. Obviously, this changes the situation from the present concern with a continuous predictor and a categorical predictor to the previous section's concern with two categorical predictors: this third strategy has already been illustrated with this particular database.

Examining the predictive power of a continuous predictor (the magnitude of *X*) within levels of a categorical predictor (observer presence) represents a strategy which Chambers, Cleveland, Kleiner, and Tukey (1983) have described as *partitioning*. This particular implementation of

Table 9.4
Hypothesis Tests Included in Meta-Analytic Data Base

Study	Statistic (*df*)	*N*	DOE	Significance		Effect Size				Year	Mag. of *X*	*OP*
				Z	*p*	Z_{FISHER}	r	r^2	d			
1	$\chi^2 = 23(1)$	110	+	4.80	8.35E−7	0.49	.457	.209	1.03	1981	7	0
2	$t = 6.5(98)$	100	+	5.91	2.08E−9	0.62	.549	.301	1.31	1984	5	0
3	$F = 15.0(1,88)$	90	+	3.71	.00010	0.40	.382	.146	0.83	1985	8	1
4	$r = .335(78)$	80	+	3.04	.00119	0.35	.335	.112	0.71	1982	5	0
5	$p = .000001(-)$	80	+	4.75	.000001	0.59	.531	.282	1.25	1983	8	0
6	$Z = 3.891(-)$	70	+	3.891	.00005	0.50	.465	.216	1.05	1988	3	1
7	$p = .01(-)$	60	+	2.33	.01	0.31	.300	.090	0.63	1990	2	1
8	$F = 10.25(1,63)$	65	+	3.07	.00107	0.39	.374	.140	0.81	1986	4	0
9	$r = .535(68)$	70	+	4.77	9.45E−7	0.60	.535	.286	1.27	1987	10	1
10	$t = 6.0(63)$	65	+	5.31	5.79E−8	0.70	.603	.364	1.51	1989	9	1

Note: DOE = Direction Of Effect (where + = expected direction,
− = unexpected direction)
Mag. of *X* = Magnitude of manipulation of independent variable *X*
OP = Observer Presence (where 1 = Present,
0 = Absent)

partitioning reveals an interesting pattern. Figure 9.1a illustrates the relation between Z for effect size and magnitude of X for the $k = 5$ hypothesis tests where the observer was absent; Figure 9.1b illustrates this relation for the $k = 5$ hypothesis tests where the observer was present.

For the $k = 5$ hypothesis tests where the observer was absent, there was a strong correlation between magnitude of X and effect size, $r = .535$. However, focused comparisons of significance levels, $Z = 1.073$, $p = .1415$, and effect sizes, $Z = 1.097$, $p = .1364$, revealed that this relation was nonsignificant. For the $k = 5$ hypothesis tests where the observer was present, the correlation between magnitude of X and effect size was stronger, $r = .683$. Focused comparisons of significance levels, $Z = 1.846$, $p = .0324$, and effect sizes, $Z = 1.666$, $p = .0478$, revealed that this relation was significant.

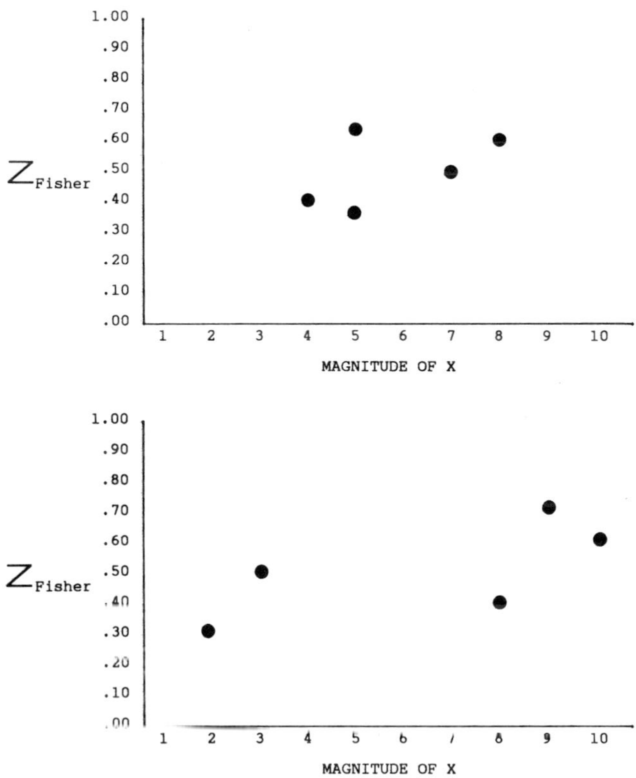

FIGS. 9.1(a) and 9.1(b). Effect size and magnitude of manipulation of X, under different conditions of observer presence (a) Observer absent, (b) Observer present.

A few things about this particular pattern of results should be made explicit. First, it should be noted that neither of the correlations between magnitude of *X* and effect size would be identified as significant using straight, primary-level inferential statistics. The probability level associated with a correlation of $r = .535$ based on 5 data points for observer absent would be $p = .1795$; the probability level associated with an $r = .683$ based on 5 data points for observer present would be $p = .1049$. This illustrates, once again, the difference between meta-analytic focused comparisons and the application of primary-level statistical techniques to meta-analytic data points, as discussed previously.

Second, consider the difference between the present results of testing the effects of a continuous predictor within levels of a categorical predictor, and the results presented before for testing the effects of one categorical predictor within levels of another categorical predictor. When magnitude of *X* was collapsed into type of *X* for the analyses reported earlier, the strength of the operationalization of *X* was shown to have no predictive power when the observer was absent, and to have only marginally significant predictive power when the observer was present. However, in the analyses where the strength of the operationalization of *X* is treated as a continuous metric, this continuous predictor is a marginal predictor of study outcomes when the observer is absent, and a significant predictor of study outcomes when the observer is present.

This parallels a tendency illustrated previously in chapters 4 and 8. At both the primary level of analysis and the meta-analytic level, treating a continuous metric as a continuous metric takes full advantage of the variability in that set of numbers; treating a continuous metric as a categorical metric loses a considerable amount of information. This does not mean that using the continuous metric, as opposed to collapsing it into a categorical metric, will always yield stronger results. If there is no predictive power to the numbers, it doesn't matter quite so much how they are scaled. However, if there is some predictive power in the numbers, that predictive power might be revealed with the continuous metric, although it might get masked in the categorical metric.

The second strategy for dealing with the complex effects of a categorical predictor and a continuous predictor is to develop a focused comparison of some arithmetic combination of the two predictors. This procedure is analogous to the one followed earlier for the combination of two categorical predictors (although it typically becomes more complex, computationally, because continuous predictors usually yield less neat λs). Once again, for example, suppose the reviewer has sound theoretical reasons to expect that the presence of the observer and the magnitude of the operationalization of X will combine in a straightforward additive manner. The meta-analytic focused comparisons that test for this type of complex effect require the derivation of λs, which capture the variability in an additive combination of the categorical predictor and the continuous predictor. Table 9.5 illustrates the derivation of these λs.

The first step is to derive the simple effect λs for the two predictors. It can be seen that, once again, the simple categorical predictor in this instance yields a simple array of +1s and −1s. However, the λs for the continuous predictor are noticeably less neat. This is because the values of the continuous predictor are unevenly spaced (e.g., the gap between successive levels of the predictor are not equivalent: The difference between 4 and 5 is not the same as the difference between 5 and 7), and because there are unequal numbers of exemplars at each level. In addition, it can be seen that these two sets of λs are not on the same metric: The contrast weights for observer presence range from −1 to +1, whereas the contrast weights for magnitude of X range from −41 to +39. So, once again, these λs are transformed into Z scores that (a) still sum to zero, (b) retain the representation of variation in the corresponding predictor, and (c) are on the same metric, having a mean of zero and a standard deviation of 1. Once these contrast weights have been transformed into Z scores, arithmetic operations can be conducted on them in order to combine them according to theoretically informative models.

For example, adding together the Zs derived from the λs produces a

Table 9.5
Derivation of λs For Additive Combination of a Categorical Predictor and a Continuous Predictor

Study	*Mag. of X*	λ_{MAG}	Z_{MAG}	OP	λ_{OP}	Z_{OP}	$Z_{MAG} + Z_{OP}$	λ_{MAG+OP}
1	7	+9	+.34	0	−1	−.95	−.61	−9.57
2	5	−11	−.41	0	−1	−.95	−1.36	−17.07
3	8	+19	+.71	1	+1	+.95	+1.66	+13.13
4	5	11	−.41	0	−1	−.95	−1.36	−17.07
5	8	+19	+.71	0	−1	−.95	−.24	−5.87
6	3	31	−1.15	1	+1	+.95	−.20	−5.47
7	2	−41	−1.53	1	+1	+.95	−.58	9.27
8	4	−21	−.78	0	−1	−.95	+1.73	+13.83
9	10	+39	+1.45	1	+1	+.95	+2.40	+20.53
10	9	+29	+1.08	1	+1	+.95	+2.03	+16.83
	Mean	0			0			
	S.D.	26.85			1.054			

new column of numbers that represents the additive combination of the magnitude of X and the presence of the observer. This new predictor is most positive when there is a strong manipulation of X in the presence of an observer, and most negative when there is a weak manipulation of X in the absence of an observer. Finally, these summed Zs are converted into a new array of λs.

When these new λs are used in focused comparisons of significance levels, $Z = 1.044$, $p = .1484$, and focused comparisons of effect sizes, $Z = 1.487$, $p = .0685$, there is a marginally suggestive indication of an additive combination of the magnitude of X and observer presence. Although this complex effect is not statistically significant, it is interesting to note that these focused comparisons produced stronger results than the additive combinations examined when magnitude of X had been collapsed into the categorical predictor of type of X. This suggests that these two predictors of observer presence and magnitude of X may somehow combine, but that an additive model of combination really does not capture the precise form of the functional combination. The reviewer might have to reconsider the theoretical accounts for the phenomenon being examined and determine whether another (perhaps more complicated) model of combination could illuminate the variability between studies.

TWO CONTINUOUS PREDICTORS

At this point, the reader will probably have already anticipated the strategies that might be followed in examining the complex effects of two continuous predictors. First, both continuous predictors could be collapsed into categorical predictors. This allows the application of the techniques of cross-blocking, testing the effects of one predictor within levels of the other predictor, and testing some arithmetic model of combination.

Second, one of the continuous predictors could be collapsed into a categorical predictor, while the other is left on its original continuous metric. This allows the testing of the effects of the continuous predictor within levels of the (now) categorical predictor, as detailed previously. This particular strategy, which gives rise to a "picture" of the effects of one continuous predictor at each level or layer or slice of another continuous predictor, has been described by Chambers et al. (1983) as a *casement display* because of its similarity to flipping through casements of fabric, tiling, or other samples in a tradesman's outlet. In addition, of course, collapsing one of the continuous predictors into a categorical predictor also allows the development of focused comparisons testing some arithmetic model of combination.

Third, one could leave each continuous predictor on its continuous metric, deriving λs for each separate predictor, and modeling some arithmetic combination, as detailed already twice before for other types of complex effects. The most informative strategy, depending on the particulars of the research domain under consideration, may very well involve some combination of these approaches.

For example, a focused comparison of an additive model that combines the two continuous predictors on their continuous metrics loses none of the potentially informative variability between studies on these two predictors. However, to illustrate and to help interpret the results of a significant additive combination focused test, the reviewer might collapse one of the predictors into a categorical predictor and present the resulting casement displays of the relation between the continuous predictor and effect size at each level of the (now) categorical predictor. Alternatively, collapsing both continuous predictors into categorical predictors, and examining the combined effects within each subsample of hypothesis tests defined by the cross-blocking procedure, can also provide an accessible and useful rendering of the combined influence of two continuous predictors.

In sum, the combined effects of two predictors can be represented, and tested, using the basic techniques of combination and focused comparison detailed in previous chapters. It should be recognized that this general strategy can be developed to apply to combinations of more than two predictors, as well as to patterns of combination other than simple, neat, linear additive models. The combined effects of multiple predictors, as well as theoretically more complex multiplicative, logarithmic, or exponential combinations, may be dictated by the issues and paradigms that are inherent in a particular research domain. However complex the theoretical combination specified by the research domain under consideration, the meta-analytic review should be able to generate adequately sophisticated tests and models of that combination.

THREE-VARIABLE RESEARCH DOMAINS

A completely different set of issues is involved in the types of complex effects that are represented in research domains that have multiple primary variables. In all of the examples considered thus far, there have been two primary variables (*X* and *Y*), as well as several theoretically interesting predictors (magnitude of *X*, observer presence) and/or practically relevant variables (year of publication). However, some research domains are devoted to the examination of phenomena that are more complex than the two-variable, *X–Y* type of issue. What if a research domain is directed toward determining whether the influence of *X* on *Y* is a direct influence, or an indirect influence through variable *Z?*

Driskell and Mullen (1989) recently had to deal with just such a complex effect at the meta-analyic level. Specifically, the phenomenon under consideration involved the interrelationships between status, expectations, and behavior. A generally consistent finding is that people respond in predictable ways to people of higher status. That link between status and behavior has been explained in many ways. One explanation, referred to as status–expectations theory, holds that the other person's status creates certain expectations that, in turn, produce the behavior differential. Other, biologically oriented, theoretical perspectives discount the role of expectations, saying that they are an epiphenomenon, and that status exerts direct effects on behavior without any intervening cognitive events. Thus, the theoretical issue is whether status influences behavior directly or indirectly through its effect on expectations.

Three strategies, illustrated in Driskell and Mullen's (1989) analyses, can be used to answer this type of complex effect theoretical question at the meta-analytic level. The first strategy requires the reviewer to derive the effect size for each study for each of the theoretically interesting hypotheses tests. For example, each of $k = 7$ hypothesis tests provided statistics representing the magnitude of the effect of status on behavior, the effect of status on expectations, and the effect of expectations on behavior. The first strategy is simply to inspect the combined magnitude for each of the effects. According to status–expectations theory, the relationship between status and behavior should not be as strong as that between status and expectations, or that between expectations and behavior. In fact, the mean weighted magnitude of effect for the relation between status and behavior, $\overline{Z}_{FISHER} = 0.39$, was weaker than the combined magnitude of effect for the relation between status and expectation, $\overline{Z}_{FISHER} = 0.65$, or than that for the relation between expectations and behavior, $\overline{Z}_{FISHER} = 0.52$.

The second strategy gives a more precise gauge of the extent to which a particular theoretical direction of influence is operating. For each study, theoretically illuminating partial correlations can be derived; the combined significance and magnitude of these theoretically illuminating partial correlations can then be determined. For example, according to status–expectations theory, the relation between expectations and behavior is an immediate and direct effect, which should not be influenced by partialling out the effects of status. On the other hand, the relation between status and behavior is an indirect effect that should be reduced to triviality when the effects of expectation are partialled out. In fact, this is generally what happened. When the effects of status were partialled out of the relation between expectation and behavior, the combined effect size, $\overline{Z}_{FISHER} = 0.38$, indicated that expectations still had a considerable direct effect on behavior. On the other hand, when the effects of expectations were partialled out of the relation between status and behavior, the combined effect sizes, $\overline{Z}_{FISHER} = 0.12$, indicated that status had little or no direct influence on behavior.

The third strategy literally provides a clear picture of this type of complex pattern of effects. The meta-analytic combinations of effect size, for both the simple direct effects and the partial correlation effects described previously, can be employed to construct what might be identified as a meta-analytic path-analysis diagram. A path analysis is generally employed at the primary level of analysis to represent potential direct and indirect effects of multiple variables in a systematic way. At the meta-analytic level, this provides a very economical summary of a tremendous amount of information regarding the interrelations between multiple primary variables. For example, Fig. 9.2 presents the path analysis rendered by Driskell and Mullen (1989) to portray the paths of effect for status, expectations, and behavior. It can be seen that the data reported in the research literature provided compelling support all along for one of the theoretical accounts for the relation between status and behavior. However, the complicated nature of the effect predicted by

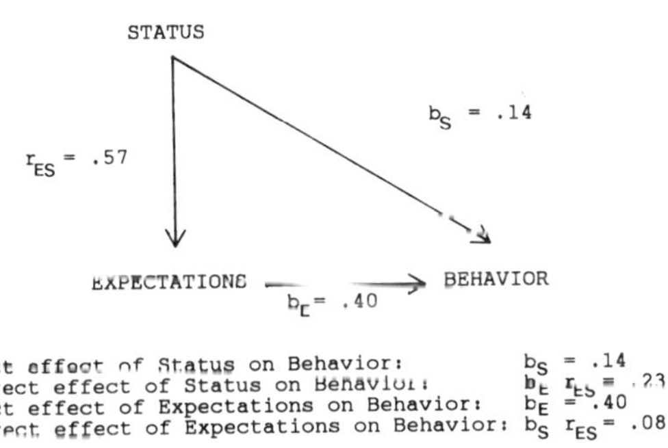

FIG. 9.2. Path analysis representing the interrelations between Status, Explanation, and Behavior (from Driskell & Mullen, 1989).

that theoretical account had, for years, precluded the explication of that support.

It is highly likely that meta-analytic integrations will begin to utilize and develop multivariate techniques, like meta-analytic path analyses and meta-analytic structural modeling, in an effort to address more complex theoretical issues at the meta-analytic level.

In sum, there are a few elegant but sophisticated strategies that the meta-analytic reviewer can follow to go beyond the basic combinations and focused comparisons in trying to predict and explain variation in study outcomes. By developing more elaborate focused comparisons, and by considering the possibility of multiple primary variables, meta-analysis can offer more and more to the enterprise of summarizing and integrating the results of primary-level research. With a careful and thoughtful understanding of these approaches to complex effects, the meta-analyst who wishes to pursue substantive theoretical issues at the meta-analytic level may find the most severe limitations to be the constraints of the research domain under consideration, and those of his or her own cleverness.

10

The Most Commonly Asked Questions About Meta-Analysis

It is clear that meta-analysis, like rock and roll, is here to stay.
—Iyengar & Greenhouse (1988)

We have considered meta-analysis at each of the steps of its implementation, from defining the hypothesis and retrieving the relevant information from each study included in the data base to addressing the three general analytic questions of central tendency, variability, and prediction and explanation. We have examined each of several meta-analytic statistical techniques and addressed a wide range of specific strategies for dealing with special cases and exceptions. At this point, we need to step back from the specifics of the steps of doing a meta-analysis and the various meta-analytic statistical techniques, and consider some broader issues about the enterprise as a whole.

In this chapter, we consider some of the more general issues that have to be examined in any implementation of meta-analysis. Aspects of some of these general issues have been implicit in our earlier discussions of specific strategies and techniques. Nonetheless, it is useful to step back and deal with these more general issues explicitly and at a broader level. This discussion can be guided by the consideration of a few general questions that are often raised about any meta-analysis. Sometimes these questions are raised in a thoughtful and sincere manner, by someone who is seriously interested in helping to improve the quality of a particular meta-analytic integration; unfortunately, however, these questions can sometimes be raised in a thoughtless and reflexive way by someone who is simply thrashing about for a damning indictment of the meta-analytic enterprise as a whole. Considering the most appropriate and defensible responses to these often-asked questions will help to highlight some of the basic issues in performing meta-analysis.

QUESTIONS ABOUT META-ANALYSIS

1. Isn't This Really Adding Apples and Oranges? In the process of integrating the results of several separate hypothesis tests, a meta-analysis necessarily involves the integration of information based on different sampling procedures, different cultural and historical configurations, and potentially different operationalizations and primary-level statistical analyses. To the extent that two hypothesis tests differ on any of these types of dimensions, the question might be raised that it is not entirely appropriate to integrate the results of these two different kinds of hypothesis tests.

This question was intricately interwoven with our consideration of the task of defining the hypothesis test in chapter 2. Some meta-analysts (e.g., Cooper, 1984) argue on behalf of casting the widest net, to include the widest range of operationalizations of independent variable and dependent variable. It should be apparent that, of the two strategies, a more narrowly focused meta-analysis is preferred by the present author. The trade-off has been highlighted before: The widest net will bring in the largest meta-analytic database, along with the potential benefits of multiple operationism. However, this widest-net approach has the potential to mask theoretically substantive patterns that are associated with unexamined or undetected differences between the various operational approaches; and, of course, it is this less selective approach that is most prone to the "apples and oranges" question. Alternatively, the narrower definition of the hypothesis to be considered preemptively thwarts the "apples and oranges" criticism and provides a more precise and specific focus on the phenomenon being considered. However, this approach brings with it the liabilities of a potentially small meta-analytic database and the restriction of generalization to a particular facet of the more general phenomenon under consideration.

Clearly, the best response to this question is to have made an explicit and defensible decision regarding the broad–narrow focus of the meta-analysis even before starting to retrieve studies, and to be prepared to examine effects separately for "apples" and for "oranges" if the broad, widest-net approach is chosen.

2. *Isn't This Data Base too Small for a Meta-Analysis?* Perhaps because some meta-analyses have been conducted with fairly large numbers of hypothesis tests (e.g., Mullen et al., 1985; Rosenthal & Rubin, 1978), some readers have developed the impression that meta-analysis can (or should) only be applied to large databases. Clearly, a larger number of hypothesis tests may increase our intuitive confidence in an analysis. However, readers of meta-analyses have to become sufficiently familiar with meta-analytic techniques to understand that the size of the database does not necessarily have anything to do with the appropriateness of the statistical techniques. As has been pointed out clearly several times (e.g., R. Rosenthal, 1983, 1984), there are meta-analytic techniques of combination and comparison of significance levels and effect sizes that are appropriate for, in fact designed for, meta-analytic databases consisting of $k = 2$ hypothesis tests. If the results of the small number of studies are sufficiently significant and strong, then the combined effects will reveal combinations of significance that are significant, and combinations of effect size that are strong. If study outcomes vary with some particular predictor to a sufficiently strong degree, the focused of comparisons of significance levels and effect sizes will reveal statistically significant results, even for a small database.

As discussed in chapter 2, one of the hidden costs of developing a more theoretically focused definition of the hypothesis test is that it sometimes elicits this question about the size of the meta-analytic database in a reflexive manner. The meta-analyst must be aware of the "catch-22" double bind here and be prepared to respond to it: If you cast the widest net in order to reap the benefits of generalizability and multiple operationism, some readers will thoughtlessly respond with "apples and oranges." If you cast a tighter net in order to reap the benefits of theoretical focus, some readers (perhaps the same ones as before) will thoughtlessly respond with "too small a database."

Another question that may be implied by a more thoughtful consideration of the "isn't it too small" question is, *Isn't this database too unrepresentative for a meta-analysis?* This is clearly a more serious problem. If there are, in fact, 100 studies available in a given research domain, and one conducts a meta-analysis on only 10 of them, one runs the risk of somehow having a nonrepresentative sampling of 10 out of the total 100; patterns discerned and conclusions drawn from this subsample of 10 may be considerably error-ridden estimates of the patterns and conclusions that one would derive from the entire population of 100 hypothesis tests.

Clearly, this potentially devastating question revolves around the adequacy of the meta-analytic reviewer's efforts to retrieve studies, as discussed at length in chapter 3. However, this serious question of the representativeness of the database is very different from the question of the size of the database. If there are, in fact, only 10 studies available in a given research domain, and one conducts a meta-analysis that includes all 10 of them, then one has a completely representative sampling (i.e., the entire population). In the extreme, a meta-analytic database can be too unrepresentative, but it cannot be too small. The meta-analytic database comprised of $k = 10$ hypothesis tests that was used throughout the book to illustrate the various analytic techniques conveyed several significant and interesting results, in spite of its small size. If, in years to follow, additional hypothesis tests lead to a meta-analytic integration that suggests a different pattern, this would not necessarily mean that the first meta-analysis was wrong. However, it might suggest that the later studies differed from the earlier studies in some interesting way, which might be pursued with meta-analytic focused comparisons in the total, larger meta-analytic database.

3. *Isn't This Research Domain too New to Have a Meta-Analysis?* In part, this question may represent a response to a small meta-analytic database, and thereby involves many of the issues raised previously. However, implicit in this particular question is the assumption that a meta-analysis can only inform the direction of future research if it based on a large number of past studies. This assumption is simply wrong.

Suppose the $k = 10$ meta-analytic database used to illustrate the various techniques throughout this book represented the sum total of all hypothesis tests conducted in a new but developing research domain (for the sake of this question, assume the year of publication for all studies was within the past 2 years). From this "too early" meta-analytic integration, the field would have learned: what the typical study outcome seems to be [and from that, a power analysis (e.g., Cohen, 1977) could estimate the sample size needed to achieve significant results in this developing research domain]; how much variability there seems to be, so far, around that typical study outcome; and, perhaps most importantly, the extent to which study outcomes do or do not vary as a function of the presence of the observer and the magnitude of the operationalization of the independent variable. Again, future meta-analyses could always be conducted to mark historical trends in study outcomes (and/or historical trends in methodology that yield changes in study outcomes at different stages in the development of the research domain).

4. *Isn't This Really "Garbage In—Garbage Out"?* The argument implied in this question, considered earlier in chapter 5, is that some or all of the research being included in the meta-analytic integration is bad research. This could mean that the hypothesis tests included in the database are of poor quality. At a more fundamental level, it might mean that the reader considers the entire domain of research to be ill-conceived and unworthy of serious consideration overall.

At one level, the meta-analytic reviewer can be faulted for having become interested in the questions pursued by a sloppy or ill-conceived research domain. In fairness to the reviewer, we should at least acknowledge that if an area of research is comprised of poorly conducted studies, this is really the fault of the primary-level researchers who did those studies, and not the reviewer (unless, of course, the reviewer also happens to have been a contributor to this problematic research domain!).

At another level, this is really a question of the definition of the hypothesis test, as detailed in chapter 2. If a body of research encompasses a wide range of poorly constructed operationalizations of the independent variable and/or the dependent variable, it may be indefensible to try to integrate these diverse hypothesis tests into a single summary. It should be recognized that in this instance, the "garbage in—garbage out" question becomes a special case of the "apples and oranges" question. In that light, it may be more defensible and more illuminating to redirect the focus of one's analyses to a smaller set of studies that use a single operationalization of the variable in question. Alternatively, one could maintain the interest in the broad, problematic research domain as a whole, but try to capture some of the diffuse variability in the literature with theoretically driven focused comparisons. Finally, one could begin to address the basic question of quality of research by careful coding for threats to internal validity, external validity, and so on, as discussed in chapter 5 (although, as we have seen, there are problems with this type of procedure as well).

It must be recognized that this question, sometimes raised in a thoughtless and careless manner, can, in particular instances, represent a devastating weakness in a meta-analysis. This was clearly articulated by Ichheiser (1949), who noted that "No refined statistical methods will be of any help if the original data upon which our theories are based are already distorted and falsified by misinterpretations operating below the level of our explicit awareness" (p. 6). Though framed in terms of primary-level research, this dictum applies directly to meta-analytic efforts just as eloquently, if not more so. The meta-analytic reviewer should carefully consider the costs, benefits, and wisdom of expending the effort and labor on the integration of a body of ill-conceived studies.

5. This Body of Research is Not Sufficiently Controversial, so Why Bother Doing a Meta-Analysis? This question assumes that the only value of doing a meta-analysis lies in resolving a controversy where one camp says that *X* affects *Y* and another camp says that *X* does not affect *Y*. Certainly, meta-analytic techniques *can* be used to resolve such controversies (e.g., Eagly & Johnson, in press). However, this question belies a caricature of meta-analysis as being simply and only a technique for obtaining average effect sizes; at this point, it should be clear that meta-analysis is much more than that. It should be recognized that occasionally an effect, which everyone assumed was there, is revealed by a meta-analysis to have not been there all along anyway (e.g., Sedikides & Ostrom, 1988). In addition, this question also belies a caricature of science as being forever at a preliminary stage of development, where researchers are trying to determine *whether X* affects *Y* and never progressing to a more sophisticated and more challenging stage of development, where we know that *X* affects *Y*, and researchers begin to consider the extent to which *X* affects *Y*, and, more importantly, *why* and *under what conditions X* affects *Y*.

This question represents a disconcerting type of intellectual arrogance, which, if occurring among college students in a psychology experiment, would be called *the hindsight bias* (Fischhoff, 1977; Slovic & Fischoff, 1977): the tendency to exaggerate one's ability to have foreseen how something would have turned out after learning the outcome. Considering the parallel between this particular question among readers of meta-analyses and the hindsight bias among subjects in experiments, it is particularly disconcerting that "debiasing instructions," which explain the occurrence of this hindsight bias to subjects in an experiment, have been shown to be generally ineffective (e.g., Slovic & Fischhoff, 1977). It remains to be seen whether the same can be said for the scenario of readers of meta-analyses being subject to the "I-knew-it-all-along" phenomenon.

6. Is This the Appropriate Analytic Strategy? This question can apply to any one of a number of choices that the meta-analyst makes in developing their meta-analytic integration: weighting study outcomes by unit 1, or by sample size, or by some indicator of study quality; treating multiple hypothesis tests from a given study as independent, or collapsing across multiple hypothesis tests to produce a single hypothesis test for each study; separate meta-analyses being conducted for each distinct operationalization, or putting all hypothesis tests together into one big pile; and so on.

As we have seen repeatedly, the meta-analytic reviewer must formulate the best decision that will be most informative regarding the phenomenon under consideration, with an effort to always make these types of decisions, and their rationale, explicit. As suggested by Fiske (1983), those researchers who wish to criticize a meta-analysis because of the choice made on one of these decisions would only have to identify the decision in question and conduct the alternative meta-analysis using the choice that they feel is more defensible for that set of studies. It is for this reason that a complete table listing the statistics and predictors for each hypothesis test included in the meta-analysis should always be presented in the write-up. Typically, whichever strategy or decision one makes, for whatever reasons, one should be prepared to confront a reader who will say, "What about doing it the other way?" (without necessarily justifying the other way as more appropriate, or without taking the tabled data and performing the alternative analyses themselves, as per Fiske's, 1983, suggestion). Therefore, the tedious but well-prepared approach to this issue would be to conduct the integration using the various alternative strategies and decisions, report the theoretically less interesting analyses anyway, and discuss and explain any discrepancies that may arise from the use of alternative strategies.

7. *Is it Appropriate to Use the Included Studies to Test Ideas That Those Studies Were Not Designed to Test?* The implication of this question is that it is somehow unfair to the original research or the original researchers to utilize those study outcomes to examine theoretically interesting ideas that may not have occurred to those involved in the original primary-level studies. A large part of the response to this question hinges on what is meant by the word "design." If design implies the intentions of the primary-level researchers (regardless of the specifics of the methodologies of the studies), then meta-analytic model testers are simply doing the same thing that narrative literary reviewers have been doing all along, by ignoring what the original primary-level researchers were designing their research to study. Narrative literary reviewers have always indulged in the reinterpretation of the results of a particular study, "designed" or intended to represent support for Theory 1, as really providing support for Theory 2. Strube and Hartman (1983) pointed to several important theoretical frameworks that have been generated from

data bases collected for other purposes (e.g., Fiedler, 1967; Zajonc, 1965), and concluded that "there is no logical reason why existing data cannot be reinterpreted" (p. 24). The major difference, once again, between the meta-analytic approach and the narrative approach is simply the rigor and the objectivity with which a meta-analytic model tester pursues this strategy.

Alternatively, if "design" is used in a more fundamental way to refer simply to methodological manipulation of some theoretically relevant predictor (regardless of the intentions of the original researchers), then the question is fallacious. Consider the $k = 10$ hypothesis tests used for illustrative purposes in this book. If hypothesis test number 7 has a magnitude of X equal to 2, and hypothesis test number 8 has a magnitude of X equal to 4, then these two hypothesis tests *are* designed to gauge the effects of differences in the magnitude of the operationalization of the independent variable (regardless of the intentions of the original researchers who conducted these primary-level studies).

One of the strengths of meta-analytic model testing is that certain research domains will involve variations across studies of predictor variables that are of a kind, or across a range, that could never be practically or ethically employed at the level of any single primary study (e.g., historical context; suicide status; racial composition of the community). Although the variation of these attributes across studies may carry with it the confound of these variables with other variables (see following), the potential confounding of relevant variables is never any less of a conceptual problem with primary research.

8. Is it Appropriate to Try to Infer Causality From Patterns Observed at the Meta-Analytic Level? This issue has been implicit in several discussions throughout earlier portions of this book, particularly in the discussion of the retrieval of predictors of study outcomes (chapter 5) and the simple and complex effects focused comparisons (chapters 8 and 9). Some sources (e.g., Cooper, 1984) unequivocally insist that the answer to this question is always no. The requirements for inferring causality are often framed as covariation, temporal sequence, and lack of spuriousness. Meta-analytic patterns are usually quite successful at satisfying the first two requirements. Correlations and focused comparisons of significance levels and effect sizes, by definition, demonstrate the extent to which study outcomes covary with the predictor under consideration. Similarly, in most instances, the predictor being coded for can be inferred to have occurred, or to have been present, before the operationalization of the dependent variable. However, lack of spuriousness can be more problematic. Lack of spuriousness specifically means that the observed association (in this context, between study outcome and predictor) cannot be attributed to some additional, third variable. Because the meta-analytic reviewer did not control the variation of the predictor across studies, the possibility exists that such an irrelevant third variable might vary across studies and thereby be the real account for variation in study outcomes across studies.

This consideration is very serious, and should be carefully considered by anyone engaged in meta-analytic model testing. However, for some types of meta-analytic predictors, the lack of spuriousness is a nonissue, and causality can be inferred much more easily. For example, recall the discussion in chapter 5 of post hoc theoretical indicators. Once again, a post hoc theoretical indicator represents some numerical facet of the study's procedure that, when subjected to some theory-driven transformation, gives rise to a new variable that can then be used to predict and explain variation in study outcomes. These post hoc theoretical indicators are theoretical transformations of an *already controlled* operationalization of an independent variable. In such an instance, spuriousness of the relation between this independent variable manipulation and study outcome applies only as much as it applies (or fails to apply) at the primary level of analysis.

For example, in the Mullen et al. (1988) integration of the effects of different components of communication network centrality (described in chapter 5), post hoc theoretical indicators were derived that represented the extent of *degree, betweenness,* and *closeness.* No third variables could be posited to covary with these post hoc theoretical predictors (which would thereby make the observed relations between predictor and study outcome a spurious relation), unless we also assumed that the manipulations of communication network structure across studies were completely unreliable and error-ridden. Given the explicit, elegant, and unequivocal features of that structural manipulation in the experimental settings included in that meta-analytic database, it becomes untenable to discount the observed meta-analytic patterns as noncausal due to the possibility of spuriousness. Thus, with post hoc theoretical indicators, spuriousness becomes a nonissue (or, at least, an issue only to the extent that it has been considered an issue at the primary level of investigation for that particular facet of the paradigm for that particular research domain).

For other types of meta-analytic predictors, spuriousness is a much more serious issue; however, it should not be considered a crippling and insurmountable issue. The possibility of spuriousness at the meta-analytic level can be addressed in a manner that is analogous to the way in which the possibility of spuriousness is addressed at the primary level. Suppose a primary-level researcher has demonstrated a significant and meaningful relation between *X* and *Y*, with the stipulation that *X* really does come before *Y*. Before concluding that *X* causes *Y*, the researcher would have to demonstrate, as best as possible, that *X* and *Y* are not themselves both the result of some third variable, *Z*. To do this, every conceivable *Z* could be measured that theoretical concerns and common sense suggest may account for the relation between *X* and *Y*. Admittedly, this researcher could never measure all possible third variables and thereby completely discount the possibility of spuriousness. However, with every additional potential third variable that is eliminated as a plausible account for the relation between *X* and *Y*, the causal effect of *X* on *Y* becomes more compelling, and the possibility of the relation between *X* and *Y* being a spurious one becomes less compelling.

By the same token, the meta-analytic reviewer who has demonstrated that study outcomes are predicted in a meaningful way by some predictor can try to account for the possibility of spuriousness. Direct codings, judges' ratings, post hoc theoretical indices, and predictors derived from archival and historical sources can all be used in an effort to determine if relevant third variables, derived from theoretical concerns and common sense, can account for the relation between study outcomes and the critical predictor. This can be accomplished using the various approaches such as cross-blocking, one predictor within levels of another predictor, and arithmetic models of combination, specified earlier. Once again, although every possible third variable cannot be measured (and therefore, spuriousness may never be completely discounted), the critical predictor derives more confidence, and the spuriousness account derives less confidence, with every potential third variable that fails to explain away the critical, interesting prediction.

Having defended the use of meta-analytic model testing against the full-bore attack of the possibility of spuriousness, it must be emphasized that the strongest case for a novel, theoretically relevant relationship discovered at the meta-analytic level is made by following up the meta-analytic model test with an examination of the relationship at the primary level of analysis, using an experimental study where spurious influences can be better controlled. However, a cautionary note, directed to both meta-analytic reviewers and readers of meta-analyses alike: At times, another type of "catch-22" double bind has developed, this time in regard to the primary-level demonstration of patterns discovered at the meta-analytic level. When a meta-analytic integration demonstrates for the first time some theoretically interesting pattern, readers and reviewers may fall into an idiomatic presentation of the question regarding the inference of causality at a meta-analytic level. Although a primary-level demonstration of the pattern discovered at the meta-analytic level is, eventually, a desirable thing, the meta-analytic discovery of the pattern should not be deemed worthless simply because it happens to be imperfect. On the other hand, when a primary-level demonstration of the pattern discovered at the meta-analytic level is finally profferred (sometimes months or years after the original meta-analytic demonstration), the primary-level demonstration may be treated as unworthy of serious scholarly consideration because it is now "old news." Both the meta-analytic discovery of a theoretically interesting pattern and the primary-level verification of that pattern under more controlled circumstances are valuable, with unique strengths and weaknesses, and should be accepted or rejected on their own merits.

GENERAL RULES

There are some simple general rules that can be distilled from this consideration of common questions and criticisms of meta-analysis. First, do everything for the right reason, and be prepared to defend it. In other words, the criteria used to select hypothesis tests into the meta-analytic database, the choice and derivation of predictors, the transformation of study outcomes to the appropriate common metrics, and the implementation of the three general analytic questions should all be carried out with an eye toward illuminating the phenomenon under consideration in the most informative and defensible manner.

Second, do everything both ways. In other words, even though the relevant theory, or common sense, dictates that one approach is more appropriate, also pursue the other approach as a supplementary analysis, and pay careful attention to the differences in results (if any). Rest assured that someone will ask to see the results of the analytic strategy you didn't report (either because such a criticism seems like an informed sort of thing to say, or because there is, in fact, a stronger case to be made in that particular research domain to examine the results that you intuited to be less informative).

Third, report everything. That is, provide a table (of the type the reader has seen repeatedly for the $k = 10$ hypothesis tests used for illustrative purposes in this book; or, of the type the reader has/will undoubtedly see(n) in meta-analyses published in mainstream scholarly journals) that lists all interesting and relevant information; present the results of the less defensible or less relevant analyses (perhaps in a footnote); and, in line with the suggestions of Fiske (1983), provide the critical reader with everything he or she would need to know if they wished to replicate your analyses or pursue alternative analyses. Even if all of this material does not make it into the final published presentation of the meta-analytic integration, it is better to provide editors and reviewers with all of the information they need to evaluate a meta-analytic integration efficiently and completely, and to later fall back upon making the supplementary (later eliminated) analyses available to interested readers upon request.

Fourth, and finally, be careful. Mistakes can be made in the various stages of a meta-analysis (defining hypothesis, collecting units of measurement, establishing common metrics, and the three general analytic questions), just as they can be made in the various stages of primary-level research (Wanous et al., 1984). Just as the primary-level researcher labors to minimize the intrusion of errors into his or her procedures, the meta-analyst should always be concerned with insuring the rigor, the meaningfulness, and the replicability of his or her efforts.

11

Advanced BASIC Meta-Analysis: An Integrated Database Management System and Statistical Package

Why is programming fun? . . . the pleasure of making things that are useful to other people. Deep within, we want others to use our work and to find it helpful. In this respect the programming system is not essentially different from the child's first clay pencil holder "for Daddy's office."

—Frederick P. Brooks (1975)

At this point, the reader has considered each of the steps involved in the development of a successful meta-analytic integration, from initially defining the hypothesis test to be considered to pursuing the more complex types of model testing at the meta-analytic level. The strengths and weaknesses of meta-analysis, as well as various prescriptive guidelines, have all been scrutinized. What remains is the actual implementation of all of these strategies and techniques.

Some phases of the meta-analytic enterprise are intellectually stimulating and gratifying on their own. The careful definition of the hypothesis can take the meta-analyst into an informative and interesting trek through theoretical and methodological areas where no one has gone before. The derivation of post hoc theoretical indices and archival or historical predictors can have about it the air of intriguing detective work. And, the thoughtful development and interpretation of model testing and theory construction at the meta-analytic level can be extremely rewarding. However, the actual numerical computations involved in performing a meta-analysis have, for me, never had any saving graces. The best thing that can be said about the tedium of repeatedly punching numbers into a calculator is that it is better than a sharp stick in the eye.

It was for this reason that we (Mullen & Rosenthal, 1985) had originally developed a software system for the computation of meta-analytic statistics. This earlier software system was comprised of 14 distinct interactive programs, which performed several of the techniques described throughout this book. However, there are many techniques described in the present book that were not represented by any programs in that earlier package. Moreover, being interactive, these simpler programs required the re-entry of the statistical data for the computation of each meta-analytic statistic.

For this reason, a "second generation" of meta-analytic software has been developed. This meta-analytic database management system and statistical package allows the user to create meta-analytic data files that contain all of the statistical information and all of the predictors for the hypothesis tests included in the meta-analytic database. These data files can be saved to disk, edited, and updated. Once a data file has been entered into the system, any one of the various meta-analytic statistical techniques can be implemented: combinations and diffuse comparisons, disjoint cluster analysis, blocking, focused comparisons examining the effects of a single predictor, effects of one predictor within levels of another predictor, combined effects of two predictors, and so on. These various meta-analytic statistical techniques can be applied to an existing data file with the push of a button (actually, at the push of two or three buttons). Finally, a meta-analytic graphics display module allows the user to generate scatterplots, frequency distributions, and funnel plots (for the database as a whole, as well as within levels of some blocking predictor).

The purpose of this chapter is to provide a brief overview to the operation of this meta-analytic database management system and statistical package. We begin by considering a few general rules of operation. Then, we consider the structure of the system as a whole, and conclude with an examination of the various tasks that the system can perform.

GENERAL RULES OF OPERATION

The database management system is designed to operate in the MS-DOS environment. The software system comprises approximately 203 K (kilobytes) of memory, and has a minimum system requirement of 256 K of RAM (random access memory). The size of the meta-analytic database that can be accommodated by this software system depends on the available RAM capacity of the user's computer system. For example, a machine with 256 K of RAM can accommodate a database of approximately 290 hypothesis tests; a machine with 384 K of RAM can accommodate approximately 1200 hypothesis tests; and, a machine with 640 K of RAM can accommodate approximately 2800 hypothesis tests.

The system is set up to accept numbers, letters, "+" signs, and "−" signs as input. Names (both predictor names and data file names) must be limited to eight characters in length. In addition, data file names may be followed by an extension (such as .BAS or .DAT), but this is not required.

The system relies on the user's own computer system's "screen dump" routine to obtain hard copies of any screen of output. Screen dump routines usually involve the simultaneous pressing of "SHIFT" and "PRINT" or "CTRL" and "P," or some similar combination of key presses. The user is advised to consult their owner's manual if they are unfamiliar with their system's screen dump routine.

STRUCTURE OF THE SYSTEM

The general structure of the data base management system and statistical package is presented in Fig. 11.1. It can be seen that the system is comprised of five distinct modules. The first module is the database management system proper, in that this is the first module one enters when calling up the system, and this module accomplishes the basic tasks of creating and handling data files and directing the user to any one of the other modules.

The next module, set off by itself in Fig. 11.1, is the module containing routines for editing existing data files. This is the only module that is not directly connected to all of the others. One can only move to the editing module from the database management system module, and one can only move to the database management system module from the editing module. Given that there is an existing data file currently in the computer's memory, the editing module allows the user to add or delete hypothesis tests, to change a particular hypothesis test, and to add or delete predictors.

The remaining three modules perform the various meta-analytic techniques we have considered throughout this book. One module performs the basic combinations and diffuse comparisons, disjoint cluster analysis, and blocking; one module performs the basic focused

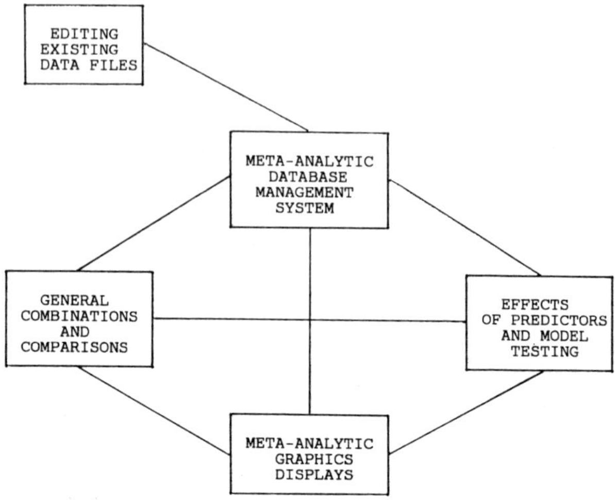

FIG. 11.1. Operation flow chart for meta-analytic database management system and statistical package.

comparisons for testing simple and complex effects of predictors; and one module performs the various meta-analytic graphics displays. We consider each of these modules in detail hereafter.

DATABASE MANAGEMENT SYSTEM

The database management system is loaded by following these steps. First, the user should turn on their computer and install their own DOS (disk operating system). At the first prompt (typically, "A>"), the user should insert the program disk, type "META" and then press the "ENTER" key. Once the program has been loaded, a title screen will appear. After pressing the "ENTER" key, the user is presented with a choice of background and text colors to be used throughout the program. If a monochrome monitor is being used, there is an indicated choice that generally presents the clearest picture on monochrome monitors. The user is then allowed to select the maximum size for the data files (i.e., k). The default value of $k = 200$ will operate on a machine with 256 K of RAM. Unless the user will have to deal with larger data files (on a machine with sufficiently larger RAM capacity), pressing the "ENTER" key will set the system to the default value of $k = 200$.

At this point, the user is prompted to remove the program disk and to insert their own data file disk. The user should have a (previously formated) disk that they devote to the storage of meta-analytic data files. Once the program has been loaded, the program disk can be removed and the data file disk can remain in the disk drive. Hard disk users can install the database management system on their hard disk and save their data files on that hard disk as well.

The menu for the database management system is presented on the screen. This menu is reproduced in Fig. 11.2.

Note that the upper-right-hand corner of every screen will display the data file currently in memory (or "NONE" if no data file is currently in memory). A data file will be in memory if a data file has just been created or if a previous data file has been loaded from the disk. In order to select one of the tasks, the user presses a letter key and then presses "ENTER."

A-CREATE DATA FILE. Upon selecting this task, the user is prompted to enter a name for the data file. Again, this data file name must be no more than eight alphanumeric characters. For example, the 10 hypothesis tests used in this book to illustrate the various analytic techniques are stored on the *Advanced* BASIC *Meta-Analysis* program disk in the data file named BOOK. The user is also promted to enter the number of hypothesis tests (not to exceed the value of k established at the beginning of the

```
ADVANCED BASIC META-ANALYSIS                DATA FILE IN MEMORY: NONE
              META-ANALYTIC DATABASE MANAGEMENT SYSTEM

SELECT TASK:
   A - CREATE DATA FILE
   B - SAVE DATA FILE TO DISK
   C - LOAD EXISTING DATA FILE FROM DISK
   D - LIST DATA FILES ON DISK
   E - DISPLAY DATA FILE IN MEMORY
   F - EDIT EXISTING DATA FILE
   G - BEGIN META-ANALYSIS:
       GENERAL COMBINATIONS AND COMPARISONS
   H - BEGIN META-ANALYSIS:
       EFFECTS OF PREDICTORS AND MODEL TESTING
   I - META-ANALYTIC GRAPHICS DISPLAYS
   J - EXIT ADVANCED BASIC META-ANALYSIS

(ENTER SELECTION)?
```

FIG. 11.2 Menu screen for database management system module.

program {default value $k = 200$}) and the number of predictors (up to 10). If any predictors are to be included in the data file, the user is then prompted to enter a name (up to eight characters) for each predictor. Then, the user is prompted to enter all of the data for each hypothesis test in the data base. This includes the type of statistic (chi-square, t test, F test, correlation coefficient, Z score or one-tailed p value), the value of the test statistic, the relevant sample size and degrees of freedom, the direction of effect, and the values of all of the predictors (if any).

Some of the restrictions for data input are standard. For example, only chi-squares on 1 degree of freedom may be entered, and only F tests with 1 degree of freedom in the numerator may be entered. Two restrictions for data input are specific to this system. First, when entering the actual numerical value of the statistic, the absolute value of the statistic should be entered, regardless of the direction of effect. In other words, if hypothesis test number 1 yielded a correlation coefficient of $r = -.75$, the user would simply enter ".75" when prompted to enter the numerical value of the statistic for hypothesis test number 1. Three steps after the entry of the numerical value of the statistic, the user is prompted to enter the direction of effect for that hypothesis test. In this instance, the user would now enter a "−", which tells the program that the previously entered ".75" was in the unexpected direction. So, the first unique restriction for data input is that absolute values of the statistics be entered, with direction of effect being represented later by a "+" or "−". Second, a special case of the foregoing restriction, only p values less than or equal to .50 may be entered into the system. A one-tailed p value of greater than .50 represents a negative direction of effect. So, once again, the "absolute" value of this particular kind of statistic must be entered, and the direction of effect specified later with a "+" or "−". Thus, p values of greater than .50 should be subtracted from 1.000 before being entered into a data file.

At the end of data entry for each hypothesis test, the user is given the option to re-enter that hypothesis test (if there happened to be an error or mistype in data entry). At the end of data entry for the data file as a whole, the system automatically saves the new data file to the disk with the name assigned at the beginning of the procedure. Data for each hypothesis test are stored in a sequential data file, in the order in which they were entered. Thus, the third hypothesis test to be entered is identified as "hypothesis test number 3." An understanding of this sequential data file structure is critical whenever the user wishes to edit the data file (see following).

B-SAVE DATA FILE TO DISK. Note that this task cannot be performed if there is no data file currently in memory. Any such attempt to perform an impossible task will result in a beep, an error message, and a return to the menu. Assuming there is a data file currently in memory, upon selecting this task the user is prompted to enter the name under which the data file in current memory will be saved. Again, this name is limited to eight characters. It is always a good idea to have an extra "back-up" disk on which one has stored all of one's data files. Note that saving the data file in current memory under a different name does not erase the old data file or change the name of the old data file; the old data file, with its old name, is still on the disk and still in current memory. This task simply saves an extra copy of the data file. If the user wishes to save an extra copy of the same data file on a back-up disk devoted to meta-analytic data files, a new (previously formated) disk can be inserted into the disk drive before pressing "B" "ENTER", and the same name can then be entered for the name of the data file. If a new disk is not inserted when the same name is entered here, it will simply write the file on top of itself (with no adverse effects, but with no extra copy being made either).

C-LOAD EXISTING DATA FILE FROM DISK. This task prompts the user to enter the name of a data file previously saved to the disk. If a back-up disk has been used, the back-up disk must be inserted before entering the name of the file. If the user enters a data file name that is not on the current disk, this will result in a beep, an error message, and a return to the menu.

D-LIST DATA FILES ON DISK. This task simply displays all of the files stored on the disk currently in the disk drive.

Note that each of the following tasks cannot be performed if there is no data file currently in memory. Selecting any one of the following tasks without a data file in memory will result in a beep, an error message, and a return to the menu.

E-DISPLAY DATA FILE IN MEMORY. This task displays all of the information about each hypothesis test in the database, one hypothesis test at a time. In addition to the statistical information and predictors entered by the user at the time of creating the file, this task also displays the common metrics of Z and one-tailed p for significance, and Z_{FISHER}, r, r^2, and d for effect size. This is extremely useful when the meta-analytic reviewer gets to the stage of writing up the report of the meta-analysis, and begins to construct the table that lists all of this information. The "screen dump" routine for the user's computer system can be used to generate a hard-copy print-out of this information.

F-EDIT EXISTING DATA FILE. This entry will take the user into the editing module.

G-BEGIN META-ANALYSIS: GENERAL COMBINATIONS AND COMPARISONS. This entry will take the user into the general combinations and comparisons module.

H-BEGIN META-ANALYSIS: EFFECTS OF PREDICTORS AND MODEL TESTING. This entry will take the user into the module that performs focused comparisons for simple and complex effects.

I-META-ANALYTIC GRAPHICS DISPLAYS. This entry will take the user into the module that generates scatterplots, frequency distributions, and funnel plots.

J-EXIT ADVANCED BASIC META-ANALYSIS. This entry will end the operation of the program, and return the user to their disk operating system.

EDITING EXISTING DATA FILES

If the user chooses "F-EDIT EXISTING DATA FILE" from the menu for the database management system, the system switches over to the editing module. The editing module menu is reproduced in Fig. 11.3.

The user should be cautious in implementing these editing routines. Once a hypothesis test or predictor is deleted, it is lost forever. The correction of a mistake made in some of these routines (noteably B and E) may require a considerable amount of tedious re-entry of data.

A-ADD HYPOTHESIS TESTS TO DATA FILE. If this editing routine is selected, the user is prompted to enter the number of hypothesis tests to be added to the data file. Then, the user is prompted to enter all of the data for these additional hypothesis tests, in exactly the same way as during the

---

```
ADVANCED BASIC META-ANALYSIS                    DATA FILE IN MEMORY: BOOK
              ROUTINES FOR EDITING EXISTING DATA FILE

SELECT EDITING TASK:
    A - ADD HYPOTHESIS TESTS TO DATA FILE
    B - DELETE HYPOTHESIS TEST FROM DATA FILE
    C - CHANGE HYPOTHESIS TEST IN DATA FILE
    D - ADD PREDICTORS TO DATA FILE
    E - DELETE PREDICTOR FROM DATA FILE
    F - EXIT DATA FILE EDIT, RETURN TO DATABASE MANAGEMENT SYSTEM

(ENTER SELECTION)?
```

FIG. 11.3 Menu screen for editing module.

data entry that occurred in the creation of the data file. These additional hypothesis tests are added at the "end" of a sequential file. So, if an additional hypothesis test were added to the $k = 10$ data base stored as BOOK, that additional hypothesis test would become the last hypothesis test, and would be identified as hypothesis test number 11. After all of the data for the additional hypothesis tests have been entered, the system will automatically save the edited data file to the disk.

B-DELETE HYPOTHESIS TEST FROM DATA FILE. If this editing routine is selected, the user is prompted to enter the number for the hypothesis test to be deleted from the data file. The system will eliminate that particular hypothesis test from the data file, each of the subsequent hypothesis tests will be "moved up" to fill in the space left by the removed hypothesis test, and the system will automatically save the edited data file to disk. Thus, if the user wished to eliminate the seventh hypothesis test from the data file BOOK ($p = .01$, etc.), they would enter "7" in response to the prompt to enter the number of the hypothesis test to be eliminated. Then, hypothesis test 7 would be eliminated, hypothesis test 8 would now become hypothesis test 7, hypothesis test 9 would become hypothesis test 8, and so on.

C-CHANGE HYPOTHESIS TEST IN DATA FILE. Upon selecting this editing routine, the user is prompted to enter the hypothesis number to be changed. Then, the information about that hypothesis test is displayed, item by item (type of statistic; value of statistic; sample size; etc.). If that particular item of information about the hypothesis test is correct, the user simply presses "ENTER" to continue on to the next bit. However, if an item has to be changed, the user simply enters the correct value before pressing "ENTER", and the new entry is substituted for the old entry. This routine is critically important if an error is uncovered in the data file after it has already been created and saved. At the end of all of the information for this particular hypothesis test, the system automatically saves the corrected data file to disk.

D-ADD PREDICTORS TO DATA FILE. Similar to the addition of hypothesis tests to the data file, this routine prompts the user to enter the number of predictors to be added to the data file (with the stipulation that the number of predictors must not exceed 10). Then, the user is prompted to enter a name for each of the new predictors. Then, the information for each of the hypothesis tests is redisplayed, and the user is prompted to enter the data for each of the new predictors for each hypothesis test in the data file. After all of the hypothesis tests have had the new predictor data added, the system automatically saves the edited data file to disk.

E-DELETE PREDICTOR FROM DATA FILE. If the user selects this editing routine, they are prompted to enter the name of the predictor to be deleted. If the user enters the name of a predictor that is not in the data file (and/or if they misspell a predictor's name), this will result in a beep, an error message, and a return to the menu. After the chosen predictor has been deleted, the system will automatically save the edited data file to disk.

F-EXIT DATA FILE EDIT, RETURN TO DATABASE MANAGEMENT SYSTEM. Upon selection of this choice, the system will return to the database management system.

GENERAL COMBINATIONS AND COMPARISONS

If the user chooses BEGIN META-ANALYSIS: GENERAL COMBINATIONS AND COMPARISONS from any of the menus, the system switches over to this analytic module. The menu for the general combinations and comparisons module is presented in Fig. 11.4.

Once again, a hard-copy print-out of the results of any of these analyses can be generated by using the "screen dump" routine for the particular computer system being used.

A-COMBINATIONS AND DIFFUSE COMPARISONS OF SIGNIFICANCE AND EFFECT SIZE. Upon selecting this option, the user is prompted to enter a weight

```
ADVANCED BASIC META-ANALYSIS                DATA FILE IN MEMORY: BOOK
              GENERAL COMBINATIONS AND COMPARISONS

SELECT ONE OF THE FOLLOWING:
   A - COMBINATIONS AND DIFFUSE COMPARISONS OF SIGNIFICANCE
       AND EFFECT SIZE
   B - BLOCKING (COMBINATIONS AND COMPARISONS WITHIN LEVELS
       OF A PREDICTOR)
   C - DISJOINT CLUSTER ANALYSIS
   D - BEGIN META-ANALYSIS:
       EFFECTS OF PREDICTORS AND MODEL TESTING
   E - META-ANALYTIC GRAPHICS DISPLAYS
   F - RETURN TO META-ANALYTIC DATABASE MANAGEMENT SYSTEM

(ENTER SELECTION)?
```

FIG. 11.4 Menu screen for general combinations and comparisons.

that will be used in combining the results of hypothesis tests. The choices are unit 1 (which will produce unweighted analyses), sample size (which is the most common weight in weighted analyses), or some predictor. If the user chooses to weight hypothesis tests by the value of some predictor, they are then prompted to enter the name of a predictor that will be used in combining results of hypothesis tests. An example of the use of a predictor would be to code studies for a predictor named "QUALITY," as described earlier, in chapter 5. Choosing to weight study outcomes by study quality might be one way to address the question of "garbage in—garbage out," as described in chapter 10. After selecting the appropriate weight, the system will generate the statistics for combination of significance (Z for combined significance, associated one-tailed p, fail-safe number for the $p = .05$ level of significance), combination of effect size ($\overline{Z}_{FISHER}$, $\bar{r}$, $\bar{r}^2$, $\bar{d}$), and the chi-squares and associated one-tailed ps for diffuse comparisons of significance and effect size.

B-BLOCKING (COMBINATIONS AND COMPARISONS WITHIN LEVELS OF A PREDICTOR). Upon selecting this option, the user is prompted to enter the name of the predictor to be used in blocking hypothesis tests into subsets. Then the user is prompted to enter the weight to be used in combining study results. The system will then generate the statistics for combination and diffuse comparison within levels of the blocking predictor.

C-DISJOINT CLUSTER ANALYSIS. Upon selecting this option, the user is prompted to select a probability level for the disjoint clusters: $p = .10$, $p = .05$, or $p = .01$. In other words, this determines the level of statistical significance that must be achieved by a gap between adjacently ranked effect sizes before that gap will be defined as a significant gap between two disjoint clusters. After selecting the level of significance to be used in this analysis, the system rank orders the hypothesis tests (from largest to smallest effect size), and displays the hypothesis test numbers and the rs for effect size within each cluster. The user is prompted to press "ENTER" between each disjoint cluster of effect sizes until all clusters have been displayed.

D-BEGIN META ANALYSIS: EFFECTS OF PREDICTORS AND MODEL TESTING. This entry will take the user into the module that performs focused comparisons for simple and complex effects.

E-META-ANALYTIC GRAPHICS DISPLAYS. This entry will take the user into the module that generates scatterplots, frequency distributions, and funnel plots.

F-RETURN TO META-ANALYTIC DATABASE MANAGEMENT SYSTEM. This entry will return the user to the database management system.

EFFECTS OF PREDICTORS AND MODEL TESTING

If the user chooses BEGIN META-ANALYSIS: EFFECTS OF PREDICTORS AND MODEL TESTING from any of the menus, the system switches over to this analytic module. The menu for the effects of predictors and model testing module is presented in Fig. 11.5.

Once again, a hard-copy print-out of the results of any of these analyses can be generated by using the "screen dump" routine for the particular computer system being used.

A-EFFECTS OF A SINGLE PREDICTOR (CORRELATIONS AND FOCUSED COMPARISONS). Upon selection of this option, the user is prompted to enter the name of the predictor to be used in a simple effect model test. The system displays three sets of analytic results. First, correlations between the predictor and each of three indices of effect size (Z_{FISHER}, *r*, and *d*) are presented. Second, the formula for the best-fitting regression line is presented for each of these three indices of effect size (Z_{FISHER}, *r*, and *d*). As discussed at length previously, these primary-level inferential statistics are not entirely appropriate for application to meta-analytic data points; however, they can serve as useful illustrations of patterns in the meta-analytic data. Finally, focused comparisons for significance levels and for effect sizes are presented, yielding *Z*s for focused comparison and corresponding one-tailed *p*s.

B-COMPARISON BETWEEN TWO LEVELS OF A BLOCKING PREDICTOR. Upon selecting this option, the user is prompted to enter the name of the blocking

---

ADVANCED BASIC META-ANALYSIS DATA FILE IN MEMORY: BOOK

EFFECTS OF PREDICTORS AND MODEL TESTING

SELECT ONE OF THE FOLLOWING:

A - EFFECTS OF A SINGLE PREDICTOR (CORRELATIONS AND FOCUSED COMPARISONS)

B - COMPARISONS BETWEEN TWO LEVELS OF A BLOCKING PREDICTOR

C - COMBINED EFFECTS OF TWO PREDICTORS

D - EFFECTS OF ONE PREDICTOR WITHIN LEVELS OF A BLOCKING PREDCICTOR

E - BEGIN META-ANALYSIS:
GENERAL COMBINATIONS AND COMPARISONS

F - META-ANALYTIC GRAPHICS DISPLAYS

G - RETURN TO META-ANALYTIC DATABASE MANAGEMENT SYSTEM

(ENTER SELECTION)?

FIG. 11.5 Menu screen for effects of predictors and model testing.

predictor under consideration, and the numerical representation of the two levels to be compared. For example, in the $k = 10$ database used for illustrative purposes throughout this book, the presence of the observer has been numerically represented as 0 (observer absent) or 1 (observer present). In this instance, the user could enter "0" and "1" (or "1" and "0"; the order does not matter). It should be recognized that categorical predictors with only two levels will yield the same focused comparison results with this selection B as they would yield with selection A (EFFECTS OF A SINGLE PREDICTOR), described previously. The unique value of this analytic routine derives from its ability to compare any two levels of a three-or-more-level categorical predictor.

C-COMBINED EFFECTS OF TWO PREDICTORS. This option prompts the user to enter the names of the two predictors and then to select a form of arithmetic combination (additive, multiplicative, or subtractive). The system then presents the correlations, regression line statistics, and focused comparison statistics, analogous to those presented in option A (EFFECTS OF SINGLE PREDICTOR). As discussed at length in chapter 9, the user should have a very clear idea of what an additive, multiplicative, or subtractive combination of the effects of the two predictors would mean before implementing these analyses.

D-EFFECTS OF ONE PREDICTOR WITHIN LEVELS OF A BLOCKING PREDICTOR. Upon selection of this option, the user is prompted to enter the name of the blocking predictor, then to enter the name of the predictor to be examined within levels of the blocking predictor. Then, the system generates the correlations, regression statistics, and focused comparisons for that second predictor, within each level of the blocking predictor.

E-BEGIN META-ANALYSIS: GENERAL COMBINATIONS AND COMPARISONS. This entry will take the user into the general combinations and comparisons module.

F-META-ANALYTIC GRAPHICS DISPLAYS. This entry will take the user into the module that generates scattergrams, frequency distributions, and funnel plots.

G-RETURN TO META-ANALYTIC DATABASE MANAGEMENT SYSTEM. This entry will return the user to the database management system.

META-ANALYTIC GRAPHICS DISPLAYS

If the user chooses META-ANALYTIC GRAPHICS DISPLAYS from any of the menus, the system switches over to this module. The menu for the meta-analytic graphics displays is presented in Fig. 11.6.

Once again, a hard-copy print-out of the results of any of these

ADVANCED BASIC META-ANALYSIS DATA FILE IN MEMORY: BOOK

META-ANALYTIC GRAPHICS DISPLAYS

SELECT ONE OF THE FOLLOWING:

A - SCATTERPLOT OF EFFECT SIZE AND PREDICTOR
B - SCATTERPLOT OF EFFECT SIZE AND PREDICTOR
(WITHIN LEVELS OF ANOTHER PREDICTOR)
C - FREQUENCY DISTRIBUTION OF EFFECT SIZE
D - FREQUENCY DISTRIBUTION OF EFFECT SIZE
(WITHIN LEVELS OF A PREDICTOR)
E - FUNNEL PLOT OF EFFECT SIZE
F - FUNNEL PLOT OF EFFECT SIZE
(WITHIN LEVELS OF A PREDICTOR)
G - BEGIN META-ANALYSIS:
GENERAL COMBINATIONS AND COMPARISONS
H - BEGIN META-ANALYSIS:
EFFECTS OF PREDICTORS AND MODEL TESTING
I - RETURN TO META-ANALYTIC DATABASE MANAGEMENT SYSTEM

(ENTER SELECTION)?

FIG. 11.6 Menu screen for meta-analytic graphics displays.

graphics displays can be generated by using the "screen dump" routine for the particular computer system being used.

A-SCATTERPLOT OF EFFECT SIZE AND PREDICTOR. Upon selecting this option, the user is prompted to select the effect size indicator (Z_{FISHER}, *r*, *d*) to be used in these scatterplots. Next, the user is prompted to enter the name of the predictor to be used in these scatterplots. Finally, the user is prompted to enter the lowest and highest value in the database for both the effect size indicator and the predictor that have been selected. The system then generates a scatterplot, formatted according to the ranges of values just specified by the user. If the user misspecifies the ranges of the effect size indicator and/or the predictor, this will result in a beep, an error message, and a return to the menu, and the user can try again (probably expanding the range of lowest and highest values, until all of the data points in the database can fit within the range specified for the scatterplot).

The user should be aware of the effects of the scaling of these scatterplots on people's implicit evaluations of bivariate relations. *Point cloud sizing,* or the relative size of the constellation of data points in a scatterplot, has been shown to influence the "perceived" correlation represented by data in a scatterplot. Cleveland, Diaconis, and McGill (1982; see also Cleveland & McGill, 1984) demonstrated how the judged association between two variables increased (by as much as 40%) as the size of the point cloud decreased relative to the frame around it.

In the context of the scatterplot routine called up by choice A, selecting extremely high and low scale values that extend well beyond the actual range of values for the predictor and effect size index will force the majority of points into a smaller region of the scatterplot, which is likely to lead viewers of this scatterplot to infer a higher degree of correlation between effect sizes and predictor than may exist in reality.

Cleveland and McGill (1984) suggest that the scale of the axes in a scatterplot extend no more than 7% to 10% beyond the highest and lowest values of the two variables. Outliers, or inordinately extreme data points, can cause problems in point cloud sizing. In order to accommodate 1 or 2 extreme values, the range of the axis may have to be extended considerably, thereby forcing the lion's share of the scatterplot into a relatively small region. Cleveland and McGill (1984) suggested that this problem could be addressed by making two scatterplots: one with the outlier(s) and one without. The user could save the data file under a different name, delete the outlying hypothesis test(s), and reproduce the scatterplot with more reasonable scaling.

B-SCATTERPLOT OF EFFECT SIZE AND PREDICTOR (WITHIN LEVELS OF ANOTHER PREDICTOR). Upon selecting this option, the user is prompted to enter the effect size indicator, the name of the blocking predictor, and the name of the predictor to be examined within levels of the blocking predictor. After the user has entered the lowest and highest values for the effect size indicator and the predictor that will be examined within levels of the blocking predictor, the system generates a scatterplot between effect size and the critical predictor within each level of the blocking predictor.

C-FREQUENCY DISTRIBUTION OF EFFECT SIZE. Upon selecting this option, the user is prompted to enter the effect size indicator to be used, as well as the lowest and highest values of this effect size indicator. The system then generates a frequency distribution for effect size.

D-FREQUENCY DISTRIBUTION OF EFFECT SIZE (WITHIN LEVELS OF A PREDICTOR). Similar to how option B provided a graphics display like option A within levels of a blocking predictor, this option yields a frequency distribution like that produced by option C, within levels of some categorical predictor. The user is prompted to enter the effect size indicator, the blocking predictor, and the lowest and highest values of the effect size indicator, and the system generates a frequency distribution for effect size for each level of the blocking predictor.

E-FUNNEL PLOT OF EFFECT SIZE. Upon selecting this option, the user is prompted to select an effect size indicator, and to enter the lowest and highest values for that effect size indicator as well as for sample sizes. The system then generates a funnel plot of effect sizes, as described in chapter 6.

F-FUNNEL PLOT OF EFFECT SIZE (WITHIN LEVELS OF A PREDICTOR). Once again, this procedure parallels the previous procedure within levels of a categorical predictor. The user is prompted to select an effect size indicator, to enter the name of a blocking predictor, and then to indicate the lowest and highest values for effect size and sample size. The system then generates a funnel plot of effect size within each level of the blocking predictor.

G-BEGIN META-ANALYSIS: GENERAL COMBINATIONS AND COMPARISONS. This entry will take the user into the general combinations and comparisons module.

H-BEGIN META-ANALYSIS: EFFECTS OF PREDICTORS AND MODEL TESTING. This entry will take the user into the module that performs focused comparisons for simple and complex effects.

I-RETURN TO META-ANALYTIC DATABASE MANAGEMENT SYSTEM. This entry will return the user to the meta-analytic database management system.

HELPFUL HINTS

Although no statistical package can be everything to everyone, this integrated meta-analytic database management system facilitates a wide range of meta-analytic statistical and graphics techniques. Note that the inevitable limitations of this software system can be overcome with a careful and creative use of the system's editing and database management system capacities. For example, suppose the user wanted to pursue more than the 10 predictors that can be accommodated by the system. The user could implement the following procedure: Create the data file with one set of 10 predictors, and save the data file twice under two different names. Then, all of the predictors can be eliminated from one of the data files, and another set of predictors can be added to the second data file, using the module for editing routines. Thus, practically speaking, any number of predictors could be accomodated by the system (10 at a time).

Similarly, the system does not accomodate missing data on the predictors. In other words, for each hypothesis test represented in the data file, there must be the full complement of predictors. This, in part, represents a prejudice on the part of the author (that all of the predictors should be ferreted out for all of the hypothesis tests before beginning the analyses). However, the user can get around this requirement by following either one of two procedures. One approach would have the user create a duplicate data file under a different name, then edit the duplicate file by eliminating the hypothesis test(s) that did not have the full complement of predictors. The other approach would have the user create a dummy variable (e.g., "MISSING", where 0 = not missing any predictor information, 1 = missing predictor information), and assigning a nonsense value (depending on the metric of the predictor in question, "0," "9999," etc.) to the theoretically interesting predictor when that information is unavailable. Then, the user can examine the effects of the predictor within levels of the dummy blocking predictor (e.g., "MISSING"). Of course, there would be no variability in the predictor under consideration when MISSING = 1, and the same nonsense variable was assigned for the predictor under consideration. The effects of the predictor when MISSING = 0 would represent the effects of the predictor among only those hypothesis tests where information regarding that predictor was available. If the latter approach were followed, the user could also test the simple effects of MISSING, as well as deriving the basic combinations of significance levels and effect sizes by blocking on the MISSING predictor. This would determine whether there were any meaningful differences between hypothesis tests where the predictor information was available and those where the predictor information was unavailable.

Although it may be more complicated to implement even more complex techniques directly, one can still use the system to examine more complicated effects. For example, the system does not generate a cross-blocking analysis of combinations of significance levels and effect sizes. However, this is easily accomplished by creating a new predictor that assigns a separate number for each cell derived from the cross-blocking procedure. Then, the application of choice B (blocking) within the general combinations and comparisons module, and the application of choice B (comparison between two levels of a blocking predictor) within the effects of predictors and model-testing module, allows the user to gauge the difference between any of the cells defined by a cross-blocking procedure.

Similarly, even more creative or complex models and combinations can be implemented with these programs. Suppose one is interested in whether a theoretically derived, complex exponential and multiplicative combination of two predictors would predict study outcomes. Although this system does not allow the user to internally derive every conceivable arithmetic combination of two predictors, the user can always derive the combination externally by following the basic procedures described in chapter 9, and then (in the editing module) add the new, unusual type of combination into the file as an added predictor (choice D in the editing module).

Thus, the present system may have to be adapted to the peculiar requirements of specific meta-analytic projects. The point is that this integrated meta-analytic database management system and statistical package can be adapted, once the user becomes familiar with the capacities of the editing and model testing modules.

As Brooks (1975) observed, "A computer program is a message from a man to a machine. The rigidly marshalled syntax and scrupulous definitions all exist to make intention clear to the dumb engine" (p. 164). It is hoped that this integrated meta-analytic database management system and statistical package is successful at both making intention clear to the dumb engine and making things that are useful to other people.

Glossary

Abstracting Services: A strategy for locating studies that are associated with key words or phrases.
Ancestry Approach: A strategy for locating earlier studies from the bibliography and references sections of already-located studies.
Blocking: The technique of disaggregating the k-included hypothesis tests into subgroups defined by levels of some categorical predictor, and then performing the basic combination and comparison procedures within each block, or level, of the categorical predictor.
Browsing: To read at random; a locating strategy that obtains studies from peripheral sources by chance exposure.
Casement Display: Examination of the predictive power of a continuous predictor within levels of another continuous predictor that has been "sliced" into several layers, or levels, on a categorical metric.
Combination of Effect Sizes: The mean effect size for the k-included hypothesis tests.
Combination of Significance Levels: The probability that the significance levels of the k-included hypothesis tests might have been obtained if the null hypothesis were true.
Conceptual Definition: A definition of a term or variable in terms of concepts.
Contrast Weight: A weight used in deriving focused test statistics, which capture the variability in the variable being tested, and which sum to equal zero.
Cross-Blocking. The technique of disaggregating the k-included hypothesis tests into subgroups defined by the levels of two orthogonal categorical predictors, and then performing the basic combination and comparison procedures within each block, or level, of the crossing of the categorical predictors.
d (Cohen's d): A number representing effect size in terms of standard deviation

units. That is, the difference between the means of two groups divided by some index of the standard deviation of the measurements involved.

Data File: A representation of a database that is stored by computer on disk.

Database: The entire constellation of statistical and predictor information for each hypothesis test included in a meta-analysis.

Dependent Variable: The measured effect, presumed to vary as a function of some specific cause (the independent variable).

Descendancy Approach: A strategy for locating studies that obtains subsequent studies that have cited earlier, already-located studies.

Diffuse Comparison of Effect Sizes: A technique that indicates the extent to which there is a significant amount of heterogeneity in effect sizes among the k-included hypothesis tests.

Diffuse Comparison of Significance Levels: A technique that indicates the extent to which there is a significant amount of heterogeneity in significance levels among the k-included hypothesis tests.

Direct Codings: Quantitative predictors that are readily available from the written report of the study.

Direction of Effect: The directional pattern of the relation between two variables (i.e., whether Y increases or decreases as a function of X). Direction of effect is often coded as being in the expected direction ("+") or in the unexpected direction ("−").

Disjoint Cluster Analysis: A technique that identifies nonoverlapping clusters of hypothesis tests, based on the effect size and sample size associated with each hypothesis test.

Effect Size: The strength or the magnitude of an effect.

Effective Reliability: An adjustment of the mean interjudge agreement as a function of the number of judges involved.

Fail-Safe Number: An estimate of the number of unretrieved studies averaging null results, or no effect, that would be needed to bring the combined results of a body of research to some specified minimum level.

File Drawer Problem: The possibility that unknown, unpublished studies might exist, whose results fail to support the pattern established by published findings.

Focused Comparison of Effect Sizes: A technique indicating the extent to which the effect sizes of the k-included hypothesis tests are significantly predictable in some systematic, specifiable manner.

Focused Comparison of Significance Levels: A technique that indicates the extent to which the significance levels of the k-included hypothesis tests are significantly predictable in some systematic, specifiable manner.

Frequency Distribution: A graphic display with frequency on the vertical axis and effect size on the horizontal axis.

Funnel Plot: A graphic display with sample size on the vertical axis and effect size on the horizontal axis. This plot should take the shape of a full, inverted funnel if there is no publication bias in operation in the research domain.

Hindsight Bias: The tendency to exaggerate one's ability to have foreseen how something would have turned out after learning the outcome.

Hypothesis Test: A primary-level statistical test of some well-defined, directional statement of the effect of one variable on another.
Independent Variable: The manipulated cause, presumed to influence some specific effect (the dependent variable).
Invisible College: The informal network of scientists working on the same problem; a strategy for locating studies that obtains studies from individual researchers who are active in the field.
Judges' Ratings: The average of numbers assigned to each hypothesis test's procedure by judges, representing the extent to which a predictor would have been involved or engaged in each hypothesis test.
k: The number of hypothesis tests in the database.
Lack of Spuriousness: The condition where the observed association between two variables (such as effect size and predictor) cannot be attributed to some additional third variable.
λ(*lambda*): A contrast weight, which can be used in focused comparisons.
Manipulation Check: A pointed question asked of subjects in experimental research, used to gauge the effectiveness of the manipulations.
Mega-Analysis: The whimsical notion of the statistical integration of the results of independent meta-analyses. Also referred to as *meta-meta-analysis* and *super-analysis.*
Meta-Analysis: The statistical integration of the results of independent studies.
Multiple Operationism: The use of many measures that are supposed to measure the same theoretical construct, but that have different error variance.
Narrative Review: The imprecise and intuitive integration of the results of independent studies.
On-Line Computer Searches: A set of strategies for locating studies that uses computerized literature databases to implement ancestry approach, descendency approach, and abstracting services.
Operational Definition: A definition of a term or variable in terms of procedures or operations of measurement or manipulation.
p (*Probability Value*): A number, ranging from 0.0 to 1.0, which represents how improbable a statistic would be if the hypothesis being tested was true.
Partitioning: The examination of the predictive power of a continuous predictor within levels of a categorical predictor.
Path Analysis: An illustration of the magnitude of potential direct and indirect influences between multiple variables.
Post hoc Theoretical Indices: Predictors derived by subjecting some numerical facet of the hypothesis test's procedures to some theory-driven transformation.
Predictor: Some attribute of the procedure, measurements, manipulations, situation, or context of hypothesis tests that may account for variability in results of the hypothesis tests.
Primary Analysis: The original statistical analysis of data by the researcher who collected the data.
Publication Bias: A bias against the publication of insignificant results.
r (*product–moment correlation coefficient*): A number representing effect size

that ranges from −1.00 (perfect inverse relation) to +1.00 (perfect direct relation).

Reliability: The extent to which measurements can be shown to have consistency or agreement.

Scatterplot: A graphic display of a correlation. Meta-analytic scatterplots usually present effect size on the vertical axis and a predictor on the horizontal axis.

Secondary Analysis: The analysis of data by someone other than the researcher who collected the data, for purposes or with analytic strategies other than those of the original researcher.

Significance Level: The degree of improbability which is deemed necessary to cast sufficient doubt upon the truth of a hypothesis to warrant its rejection.

Validity: The extent to which measurements can be shown to measure what they are supposed to measure.

Z (*Standard Normal Deviate*): A standardized metric that has a mean of zero and a variance of one.

Z_{FISHER} (*Fisher's Z transformation of r*): A standardized metric representing effect size that makes the sampling distribution of *r* approximately Gaussian.

References

Almeida, M. C., & Denham, S. A. (1984, April). *Interpersonal cognitive problem-solving: A meta-analysis.* Paper presented at the 55th Annual Meeting of the Eastern Psychological Association, Baltimore, MD.

Andrews, G., & Harvey, R. (1981). Does psychotherapy benefit neurotic patients? *Archives of General Psychiatry, 38,* 1203–1208.

Arvey, R. D., & Campion, J. E. (1982). The employment interview: A summary of recent research. *Personnel Psychology, 35,* 736–765.

Atkinson, D. R., Furlong, M. J., & Wampold, B. E. (1982). Statistical significance, reviewer evaluations, and the scientific process: Is there a (statistically) significant relationship? *Journal of Counseling Psychology, 29,* 189–194.

Atkinson, S. D., & Dolan, D. R. (1983). In search of research studies in on-line data bases. *Online, 7,* 51–63.

Bakan, D. (1967). *On method.* San Francisco: Jossey-Bass.

Bangert-Drowns, R. L. (1986). Review of developments in meta-analytic method. *Psychological Bulletin, 99* (3), 388–399.

Baum, A., Aiello, J. R., & Calesnick, L. E. (1978). Crowding and personal control: Social density and the development of learned helplessness. *Journal of Personality and Social Psychology, 36* 1000–1011.

Becker, B. J. (1987). Applying tests of combined significance in meta-analysis. *Psychological Bulletin, 102* (1), 164–171.

Bracey, G. W. (1982). Computers in education: What the research shows. *Electronic Learning, 2,* 51–54.

Brooks, F. P. (1975). *The mythical man-month: Essays on software engineering.* Reading, MA: Addison-Wesley.

Bullock, R. J., & Svyantek, D. J. (1985). Analyzing meta-analysis: Potential problems, an unsuccessful replication, and evaluation criteria. *Journal of Applied Psychology, 70* (1), 108–115.

Campbell, D., & Stanley, J. (1963). *Experimental and quasi-experimental designs for research.* Chicago: Rand McNally.

Carlson, M., & Miller, N. (1987). Explanation of the relation between negative mood and helping. *Psychological Bulletin, 102* (1), 91–108.

Chambers, J. M., Cleveland, W. S., Kleiner, B., & Turkey, P. A. (1983). *Graphic methods for data analysis.* Belmont, CA: Wadsworth.

Chase, L. J., & Chase, R. B. (1976). A statistical power analysis of applied psychological research. *Journal of Applied Psychology, 61,* 234–237.

Chow, S. L. (1988). Significance test or effect size? *Psychological Bulletin, 103* (1), 105–110.

Churchill, G. A., Ford, N. M., Hartley, S. W., & Walker, O. C. (1985). The determnants of salesperson performance: A meta-analysis. *Journal of Marketing Research, 22,* 103–118.

Cicchetti, D., & Eron, L. (1979). The reliability of manuscript reviewing for the *Journal of Abnormal Psychology. Proceedings of the American Statistical Association* (Social Statistics Section), *2,* 596–600.

Clagett, G. P., & Reisch, J. S. (1988). Prevention of venons thromboembolism in general surgical patients: Results of meta-analysis. *Anals of Surgery 208,* 227–240.

Cleveland, W. S., Diaconis, P., & McGill, R. (1982). Variables on scatterplots look more highly correlated when the scales are increased. *Science, 216,* 1138–1141.

Cleveland, W. S., & McGill, R. (1984). The many faces of a scatterplot. *Journal of the American Statistical Association, 79* (388), 807–822.

Cohen, J. (1977). *Statistical power analysis for the behavioral sciences.* New York: Academic Press.

Cohen, P. A., Kulik, J. A., & Kulik, C. (1982). Education outcomes of tutoring: A meta-analysis of findings. *American Educational Research Journal, 14,* 237–248.

Cook, T., & Campbell, D. (1979). *Quasi-experimentation.* Chicago: Rand McNally.

Cook, T. D., & Leviton, L. C. (1980). Reviewing the literature: A comparison of traditional methods with meta-analysis. *Journal of Personality, 48,* 449–472.

Cooper, H. M. (1979). Statistically combining independent studies: A meta-analysis of sex differences in conformity research. *Journal of Personality and Social Psychology, 37,* 131–135.

Cooper, H. M. (1982). Scientific guidelines for conducting integrative research reviews. *Review of Educational Research, 52,* 291–302.

Cooper, H. M. (1983). Methodological determinants of outcomes of synthesis of research literature. In P. Wortman (Chair), *An analysis of methodologies used in synthesizing research on desegregation and student achievement.* Symposium conducted at the annual meeting of the American Educational Research Association, Montreal.

Cooper, H. M. (1984). *The integrative research review: A social science approach.* Beverly Hills, CA: Sage.

Cooper, H. M., & Rosenthal, R. (1980). Statistical versus traditional procedures for summarizing research findings. *Psychological Bulletin, 87,* 442–449.

Cowles, M., & Davis, C. (1982). On the origins of the .05 level of statistical significance. *American Psychologist, 37* (5), 553–558.

Darlington, R. B. (1970). Is kurtosis really "peakedness"? *The American Statistician, 24,* 19–22.

DePaulo, B. M., Zuckerman, M., & Rosenthal, R. (1980). Detecting deception: Modality effects. In L. Wheeler (Ed.), *Review of personality and social psychology* (pp. 125–162). New York: Sage.

Dillon, J. T. (1982). Superanalysis. *Evaluation News, 3* (4), 35–43.

Driskell, J. E., & Mullen, B. (1989). *Status, expectations, and behavior: A meta-analytic review and test of the theory*. Unpublished manuscript, Syracuse University, Syracuse, NY.

Dusek, J. B., & Joseph, G. (1983). The bases of teacher expectancies: A meta-analysis. *Journal of Educational Psychology, 75,* 327–346.

Eagly, A. H. (1978). Sex differences in influenceability. *Psychological Bulletin, 85,* 86–116.

Eagly, A. H. (1986). Some meta-analytic approaches to examining the validity of gender-differences research. In J. S. Hyde & M. C. Linn (Eds.), *The psychology of gender: Advances through meta-analysis* (pp. 159–177). Baltimore, MD: Johns Hopkins University Press.

Eagly, A. H., & Carli, L. L. (1981). Sex of researchers and sex-typed communications as determinants of sex differences in influenceability: A meta-analysis of social influence studies. *Psychological Bulletin, 90,* 1–20.

Eagly, A. H., & Crowley, M. (1986). Gender and helping behavior: A meta-analytic review of the social psychological literature. *Psychological Bulletin, 100* (3), 283–308.

Eagly, A. H., & Johnson, B. T. (in press). Gender and leadership style: A meta-analysis. *Psychological Bulletin.*

Eagly, A. H., & Steffen, V. J. (1986). Gender and aggressive behavior: A meta-analytic review of the social psychological literature. *Psychological Bulletin, 100* (3), 309–330.

Eysenck, J. J. (1978). An exercise in mega-silliness. *American Psychologist, 33,* 517.

Fagley, N. S., & McKinney, I. J. (1983). Reviewer bias for statistically significant results: A reexamination. *Journal of Counseling Psychology, 30* (2), 298–300.

Fiedler, F. E. (1967). *A theory of leadership effectiveness.* New York. McGraw Hill.

Fischhoff, B. (1977). Perceived informativeness of facts. *Journal of Experimental Psychology: Human Perception and Performance, 3,* 349–358.

Fisher, R. A. (1932). *Statistical methods for research workers* (4th ed.). London: Oliver and Boyd.

Fisher, R. A. (1938). *Statistical methods for research workers* (7th ed.). London: Oliver and Boyd.

Fiske, D. W. (1983). The meta-analytic revolution in outcome research. *Journal of Consulting and Clinical Psychology, 51,* 65–70.

Flubert, S. (1976). Driver and pedestrian characteristics. In J. E. Baerwald (Ed.), *Transportation and traffic engineering handbook.* Englewood Cliffs, NJ: Prentice-Hall.

Gibbons, J. D., Olkin, I., & Sobel, M. (1977). *Selecting and ordering populations: A new statistical methodology.* New York: Wiley.

Glass, G. (1976). Primary, secondary and meta-analysis of research. *Educational Research, 5,* 3–8.

Glass, G. (1977). Integrating findings: The meta-analysis of research. *Review of Research in Education, 5,* 351–379.

Glass, G. (1980). Summarizing effect sizes. In R. Rosenthal (Ed.), *New directions for methodology of social and behavioral science: Quantitative assessment of research domains.* San Francisco: Jossey-Bass.

Glass, G., McGraw, B., & Smith, M. L. (1981). *Meta-analysis in social research.* Beverly Hills, CA: Sage.

Glass, G., & Smith, M. L. (1978). Reply to Eysenck. *American Psychologist, 33* 517–518.

Glass, G., & Smith, M. L. (1979). Meta-analysis of research on the relationship of class-size and achievement. *Educational Evaluation and Policy Analysis, 1,* 2–16.

Gottfredson, S. (1978). Evaluating psychological research reports. *American Psychologist, 33* 920–934.

Greenwald, A. G. (1975). Consequences of prejudice against the null hypothesis. *Psychological Bulletin, 82,* 1–20.

Hall, J. A. (1978). Gender effects in decoding nonverbal cues. *Psychological Bulletin, 85,* 845–857.

Hall, J. A. (1980). Gender differences in nonverbal communication skills. In R. Rosenthal (Ed.), *New directions for methodology of social and behavioral science: Quantitative assessment of research domains* (pp. 63–78). San Francisco: Jossey-Bass.

Hedges, L. V., & Olkin, I. (1983). Clustering estimates of effect magnitude from independent studies. *Psychological Bulletin, 93,* 563–573.

Hedges, L. V., & Olkin, I. (1985). *Statistical methods for meta-analysis.* Orlando, FL: Academic Press.

Hembree, R., & Dessart, D. J. (1986). Effects of handheld calculators in precollege mathematics education: A meta-analysis. *Journal for Research in Mathematics Education, 17,* 83–99.

Hildebrand, D. K. (1971). "Kurtosis measures bimodality?" *The American Statistician, 25* (1), 42–43.

Huff, D. (1954). *How to lie with statistics.* New York: Norton.

Hunter, J. E., Schmidt, F. L., & Hunter, R. (1979). Differential validity of employment tests by race: A comprehensive review and analysis. *Psychological Bulletin, 86,* 721–735.

Ichheiser, G. (1949). Misunderstandings in human relations: A study in false perception. *American Journal of Sociology, 55* (2), 1–70.

Israel, M. (1981). A meta-analysis of the Los Angeles Unified School District's efforts in achieving its goals of desegregation: Improved academic achievement and race relations. *Dissertation Abstracts International, 41,* 11-A, 4651–4652.

Iyengar, S., & Greenhouse, J. B. (1988). Selection models and the file drawer problem. *Statistical Science, 3* (1), 109–135.

James, L. R., Demaree, R. G., & Mulaik, S. A. (1986). A note on validity generalization procedures. *Journal of Applied Psychology, 71* (3), 440–450.

Johnson, B. T. (1989). *DSTAT.* Unpublished software package, Syracuse University, Syracuse, NY.

Kamin, L. J. (1978). Comment on Munsinger's review of adoption studies. *Psychological Bulletin, 85,* 194–201.

Kazdin, A., Durac, J., & Agteros, T. (1979). Meta-meta-analysis: A new method for evaluating therapy outcome. *Behavioral Research and Therapy, 17,* 397–399.

Kerlinger, F. N., & Pedhauzer, E. J. (1973). *Multiple regression in behavioral research.* New York: Holt, Rinehart, and Winston.

Kirk, R. E. (1982). *Experimental design: Procedures for the behavioral sciences.* Monterey: Brooks/Cole.

Latane, B. (1981). The psychology of social impact. *American Psychologist, 36,* 343–356.

LeBon, G. (1895/1960). *The crowd.* New York: Viking.

Light, R. J., & Pillemer, D. B. (1984). *Summing up: The science of reviewing research.* Cambridge, MA: Harvard University Press.

Longnecker, M. P., Berlin, J. A., Orza, M. J., & Chalmers, T. C. (1980). A meta-analysis of alcohol consumption in relation to risk of breast cancer. *Journal of the American Medical Association, 260,* 652–656.

Lord, R. G., deVader, C. L., & Alliger, G. M., (1986). A meta-analysis of the relation between personality traits and leadership perceptions: An application of validity generalization procedures. *Journal of Applied Psychology, 71* (3), 402–410.

Lush, J. L. (1931). Predicting gains in feeder cattle and pigs. *Journal of Agricultural Research, 42,* 853–881.

Maccoby, E., & Jacklin, C. (1974). *The psychology of sex differences.* Stanford: Stanford University Press.

Marks, G., & Miller, N. (1987). Ten years of research on the false-consensus effect: An empirical and theoretical review. *Psychological Bulletin, 102* (1), 72–90.

Marsh, H., & Ball, S. (1981). Interjudgemental reliability of reviews for the *Journal of Educational Psychology, Journal of Educational Psychology, 73,* 872–880.

Masters, J. C., Cantor, N. E., & Hedges, L. V. (1988). Editorial. *Psychological Bulletin, 103* (1), 3–4.

Miller, N. (1988 August). *Meta-analytic model-testing in social psychology.* Symposium presented at the Annual Meeting of the American Psychological Association, Atlanta, GA.

Miller, N., & Davidson-Podgorny, G. (1987). Theoretical models of intergroup relations and the use of cooperative teams as an intervention for desegregated settings. In C. Hendrick (Ed.), *Group processes and intergroup relations, Review of Personality and Social Psychology* (pp. 41–67). New York: Spye.

Mintz, J. (1983). Integrating research evidence: A commentary on meta-analysis. *Journal of Consulting and Clinical Psychology, 51,* 71–75.

Moors, J. J. A. (1986). The meaning of kurtosis: Darlington reexamined. *The American Statistician, 82,* 623–659.

Mullen, B. (1985). Strength and immediacy of sources: A meta-analytic evaluation of the forgotten elements of social impact theory. *Journal of Personality and Social Psychology, 48* (6), 1458–1466.

Mullen, B., Atkins, J. L., Champion, D. S., Edwards, C., Hardy, D., Story, J. E., & Vanderklok, M. (1985). The false consensus effect: A meta-analysis of 155 hypothesis tests. *Journal of Experimental Social Psychology, 21,* 262–283.

Mullen, B., Copper, C., & Driskell, J. E. (in press). Jaywalking as a function of model behavior. *Personality and Social Psychology Bulletin.*

Mullen, B., & Felleman, V. (in press). Tripling in the dorms: A meta-analytic integration. *Basic and Applied Social Psychology.*

Mullen, B., & Hu, L. (1988). Social projection as a function of cognitive mechanisms: Two meta-analytic integrations. *British Journal of Social Psychology, 27,* 333–356.

Mullen, B., & Johnson, C. (in press). Distinctiveness-based illusory correlations and stereotyping: A meta-analytic integration. *British Journal of Social Psychology.*

Mullen, B., Johnson, C., & Salas, E. (1989). *Productivity loss in brainstorming groups: A meta-analytic integration and different solution to the riddle.* Unpublished manuscript, Syracuse University, Syracuse, NY.

Mullen, B., & Riordan, C. (1988). Self-serving attributions for performance in naturalistic settings: A meta-analytic review. *Journal of Applied Social Psychology, 18* (1), 3–22.

Mullen, B., & Rosenthal, R. (1985). *BASIC meta-analysis: Procedures and programs.* Hillsdale, NJ: Lawrence Erlbaum Associates.

Mullen, B., Salas, E., & Driskell, J. E. (1989). *Salience, motivation, and artifact as contributions to the relation between participation rate and leadership.* Unpublished manuscript, Syracuse University, Syracuse, NY.

Mullen, B., Salas, E., & Johnson, C. (1988, August). *Components of positional centrality in communication networks: A meta-analytic integration.* Paper presented at the 24th International Congress of Psychology, Sydney, Australia.

Munsinger, H. (1974). The adopted child's IQ: A critical review. *Psychological Bulletin, 82,* 623–659.

Oldridge, N. B., Guyatt, G. H., Fischer, M. E., & Rimm, A. A. (1980). Cardiac rehabilitation after myocardial in Farction: Combined Experience of randomized clinical trials. *Journal of the American Medical Association, 260,* 945–950.

Oliver, L. W., & Spokane, A. R. (1983). Research integration: Approaches, problems, and recommendations for research reporting. *Journal of Counseling Psychology, 30* (2), 252–257.

Orwin, R. G. (1983). A fail-safe *N* for effect size. *Journal of Educational Statistics, 8,* 157–159.

Overall, J. E., & Rhoades, H. M. (1986). Beware of a half-tailed test. *Psychological Bulletin, 100* (1), 121–122.

Pearson, K. (1933a). Appendix to Dr. Elderton's paper on "The Lanarkshire milk experiment." *Annals of Eugenics, 5,* 337–338.

Pearson, K. (1933b). On a method of determining whether a sample size *n* supposed to have been drawn from a parent population having a known probability integral has probably been drawn at random. *Biometrika, 25,* 379–410.

Petty, R. E., & Cacioppo, J. T. (1981). *Attitudes and persuasion: Classic and contemporary approaches.* Dubuque, IA: Brown.

Pillemer, D. B., & Light, R. J. (1980). Benefiting from variation in study outcomes. In R. Rosenthal (Ed.), *New directions for methodology of social and behavioral science: Quantitative assessment of research domains* (pp. 1–11). San Francisco: Jossey-Bass.

Platt, J. R. (1964). Strong inference. *Science, 146,* 347–352.

Price, D. (1966). Collaboration in an invisible college. *American Psychologist, 21,* 1011–1018.

Rachman, S. J., & Wilson, G. T. (1980). *The effects of psychological therapy: Second enlarged edition.* New York: Pergamon Press.

Rosenthal, M. (1984). *Bibliographic retrieval for the social and behavioral science meta-analyst.* Unpublished manuscript.

Rosenthal, R. (1961, August). On the social psychology of the psychological experiment: With particular reference to experimenter bias. In H. W. Riecken (Chair), *On the social psychology of the psychological experiment.* Symposium conducted at the meeting of the American Psychological Association, New York.

Rosenthal, R. (1963). On the social psychology of the psychological experiment: The experimenter's hypothesis as unintended determinant of experimental results. *American Scientist, 51,* 268–283.

Rosenthal, R. (1964). Effects of the experimenter on the results of psychological research. In B. A. Maher (Ed.), *Progress in experimental personality research, 1,* (pp. 79–114). New York: Academic Press.

Rosenthal, R. (1966). *Experimenter effects in behavioral research.* (rev. ed., 1976). New York: Appleton-Century-Crofts.

Rosenthal, R. (1968). Experimenter expectancy and the reassuring nature of the null hypothesis decision procedure. *Psychological Bulletin Monograph Supplement, 70,* 30–47.

Rosenthal, R. (1969). Interpersonal expectations. In R. Rosenthal & R. L. Rosnow (Eds.), *Artifact in behavioral research* (pp. 181–277). New York: Academic Press.

Rosenthal, R. (1979). The "file drawer problem" and tolerance for null results. *Psychological Bulletin, 86,* 638–641.

Rosenthal, R. (1983). Assessing the statistical and social importance of the effect of psychotherapy. *Journal of Consulting and Clinical Psychology, 51,* 4–13.

Rosenthal, R. (1984). *Meta-analytic procedures for social research.* Beverly Hills, CA: Sage.

Rosenthal, R., & Hall, J. (1981). Critical values of *Z* for combining independent probabilities. *Replications in Social Psychology, 1* (2), 1–6.

Rosenthal, R., & Rosnow, R. L. (1975). *The volunteer subject.* New York: John Wiley.

Rosenthal, R., & Rosnow, R. L. (1984). *Essentials of behavioral research: Methods and data analysis.* New York: McGraw-Hill.

Rosenthal, R., & Rosnow, R. L. (1985). *Contrast analysis: Focused comparisons in the analysis of variance.* New York: Cambridge University Press.

Rosenthal, R., & Rubin, D. B. (1971). Pygmalion reaffirmed. In J. D. Elashoff & R. E. Snow (Eds.), *Pygmalion reconsidered.* Worthington, OH: Charles A. Jonis.

Rosenthal, R., & Rubin, D. B. (1978). Interpersonal expectancy effects: The first 345 studies. *The Behavioral and Brain Sciences, 3,* 410–415.

Rosenthal, R., & Rubin, D. B. (1986). Meta-analytic procedures for combining studies with multiple effect sizes. *Psychological Bulletin, 99* (3), 400–406.

Rosenthal, R., & Rubin D. B. (1988). Comment: Assumptions and procedures in the file drawer problem. *Statistical Science, 3* (1), 120–125.

Rosnow, R., & Suls, J. (1970). Reactive effects of pretesting in attitude research. *Journal of Personality and Social Psychology, 15,* 338–343.

Royall, R. M. (1986). The effect of sample size on the meaning of significance tests. *The American Statistician, 40* (4), 313–315.

Sanders, G. S., & Mullen, B. (1983). Accuracy in perceptions of consensus: Differential tendencies of people with majority and minority positions. *European Journal of Social Psychology, 13,* 57–70.

Scarr, S., & Weber, B. (1978). The reliability of reviews for the *American Psychologist. American Psychologist, 33,* 935.

Schmidt, F. L., Hunter, J. E., & Pearlman, K. (1981). Task differences as moderators of aptitude test validity in selection: A red herring. *Journal of Applied Psychology, 66,* 166–185.

Sedikides, C., & Ostrom, T. M. (1988). Are person categories used when organizing information about unfamiliar sets of persons? *Social Cognition, 6* (3), 252–267.

Shapiro, D. A., & Shapiro, D. (1983). Comparative therapy outcome research: Methodological implications of meta-analysis. *Journal of Consulting and Clinical Psychology, 51,* 42–53.

Shaw, M. E. (1964). Communication networks. In L. Berkowitz (Ed.), *Advances in experimental social psychology* (vol. 1, pp. 111–147). New York: Academic Press.

Siegel, S. (1956). *Nonparametric methods for the behavioral sciences.* New York: McGraw-Hill.

Siegel, S., & Castellan, N. J., Jr. (1988). *Nonparametric statistics for the behavioral sciences* (2nd ed.). New York: McGraw-Hill.

Slovic, P., & Fischhoff, B. (1977). On the psychology of experimental surprises. *Journal of Experimental Psychology: Human Perception and Performance, 3,* 544–551.

Smart, R. G. (1964). The importance of negative results in psychological research. *Canadian Psychologist, 52,* 225–232.

Smith, M. L. (1980). Integrating studies of psychotherapy outcomes. In R. Rosenthal (Ed.), *New directions for methodology of social and behavioral science: Quantitative assessment of research domains* (pp. 47–61). San Francisco: Jossey-Bass.

Smith, M. L., & Glass, G. (1977). Meta-analysis of psychotherapy outcomes. *American Psychologist, 32,* 752–760.

Smith, N. L. (1982). Evaluative applications of meta- and mega-analysis. *Evaluative News, 3* (4), 43–47.

Snedecor, G. W. (1946). *Statistical methods* (4th ed.). Ames: Iowa State College Press.

Sterling, T. D. (1959). Publication decisions and their possible effects on interfaces drawn from tests of significance—or vice versa. *Journal of the American Statistical Association, 54,* 30–34.

Stock, W. A., Okun, M. A., Haring, M. J., Miller, W., Kinney, C., & Ceurvorst, R. W. (1982). Rigor in data synthesis: A case study of reliability in meta-analysis. *Educational Researcher, 11,* 10–14.

Stouffer, S. A. (1949). *The American soldier (Vol. 1: Adjustment during army life).* Princeton: Princeton University Press.

Strube, M. J., & Garcia, J. E. (1981). A meta-analytic investigation of Fieldler's contingency model of leadership effectiveness. *Psychological Bulletin, 90,* 307–321.

Strube, M. J., & Hartman, D. P. (1983). Meta-analysis: Techniques, applications, and functions. *Journal of Consulting and Clinical Psychology, 51,* 14–27.

Suls, J., & Fletcher, B. (1985). The relative efficacy of avoidant and nonavoidant coping strategies: A meta-analysis. *Health Psychology, 4* (3), 249–288.

Thorndike, R. L. (1933). The effect of the interval between test and retest on the constancy of the IQ. *Journal of Educational Psychology, 24,* 543–549.

Tran, Z. V., & Weltman, A. (1985). Differential effects of exercise on serum lipid and lipoprotein levels seen with changes in body weight: A meta-analysis. *Journal of the American Medical Association, 254,* 919–924.

Wachter, K. W. (1988). Disturbed by Meta-analysis? *Science, 241,* 1407–1408.

Walker, H. M., & Lev, J. (1953). *Statistical inference.* New York: Holt, Rinehart, and Winston.

Wanous, J. P., Sullivan, S. E., & Malinak, J. (1989). The role of judgment calls in meta-analysis. *Journal of Applied Psychology, 74,* 259–264.

Webb, E., Campbell, D., Schwartz, R., Sechrest, L., & Grove, J. (1981). *Nonreactive measures in the social sciences.* Boston: Houghton Mifflin.

White, K. R. (1982). The relation between socioeconomic status and academic achievement. *Psychological Bulletin, 91,* 461–481.

Wilson, G. T., & Rachman, S. J. (1983). Meta-analysis and the evaluation of psychotherapy outcome: Limitations and liabilities. *Journal of Consulting and Clinical Psychology, 51,* 54–64.

Zajonc, R. (1965). Social facilitation. *Science, 149,* 269–274.

Author Index

Z

Subject Index